Benjamin Ward Richardson

**Biological Experimentation**

Its Function and Limits

Benjamin Ward Richardson

**Biological Experimentation**
*Its Function and Limits*

ISBN/EAN: 9783337366926

Printed in Europe, USA, Canada, Australia, Japan

Cover: Foto ©berggeist007 / pixelio.de

More available books at **www.hansebooks.com**

# BIOLOGICAL EXPERIMENTATION

## Its Function and Limits

*INCLUDING ANSWERS TO NINE QUESTIONS*
*SUBMITTED FROM THE*

## LEIGH-BROWNE TRUST

BY

SIR BENJAMIN WARD RICHARDSON
M.D., F.R.S.

LONDON
GEORGE BELL & SONS
YORK STREET, COVENT GARDEN
1896

# CONTENTS.

# INTRODUCTION.

SOME years ago the council of the Leigh-Browne Trust invited me to give a clear and open reply to the nine questions stated in this little work. They knew the position in which I stood in regard to experimentation, and preferred that one, who has defended experiment, should answer for it rather than an opponent to that method of research. I have replied after the manner in which the subject now presents itself to my own mind. If I have said anything contrary to previous views I may have expressed, the reader must pardon me, for I die daily as I learn ; if I have said things to him not strong enough against experimentation, he must forgive me, as not being an advocate on his side, nor charged with his condemning fire.

Since the work commenced a fierce controversial fight has raged on some of the points discussed ; but considering furious controversy as fatal to the proper settlement of any question I took no part in it ; and I have held back from publication that a more tranquil state of the public mind on a great public question may be considered apart from anger. It must be admitted by all persons of fair judgment that the forces arrayed on each side are governed by proper intentions with perhaps too much of enthusiasm on each side. I have often felt that experimentalists are too blindly enthusiastic, and forget too readily what other people deeply feel and resent, in the hope of doing something that may be generally useful, a hope which is too often apt to deceive. On the other hand, I am sure their violent opponents constantly misinterpret them from misunderstanding their intention. Both sides are led by feeling, or, more correctly stated, by sentiment, while each side assumes that it is led by argument ; the one is feeling for another's pain, the other for what really is, by reflection, its own.

In accepting the task of answering the

questions submitted to me, I assume, therefore, the fullest liberty of expression. I am not bound personally to particular views on the matter in hand, but report, as invited, from my experience how far biological science, in its special application to the investigation of the nature, cause, prevention, and cure of disease, can progress without the infliction of pain, and how it can open up new fields of research less the imposition of suffering on the creation.

I can have no objection to answer the questions submitted to me, and in the answers will be as fair as possible. I shall use my own past work in respect to experimentation, and shall answer the questions apart from what I may conceive to be the wishes of the questioners themselves.

If in creation there were no pain, if no pain could be extorted except by a physiologist, a physiologist inflicting pain, even for the cure of disease or the reduction of death-rates, would be an accepted criminal by the general voice of mankind. But Nature is a laboratory of pain on the most gigantic scale ; she stands at nothing in the way of infliction ; spares nothing that is sentient. She inflicts pain for her own purposes, and she keeps it going. We call the pain of nature the cross to which we are subjected. But some argue that in endeavouring to extinguish the pain of Nature, they are justified when, in their designs, they follow hers in this particular, whereby they set up another cross. I may therefore study the subject before me in this essay as a *crux crucis*, in which fancy I take pain to be the crux of Nature, and the cross of the cross the pain which man, for any reason, may add to the pain of Nature.

If we think of the pain Nature inflicts it seems far more cruel than any experimentalist can possibly conceive. If a man could impose on one sex the same pain as Nature imposes on the female sex during childbirth he would be a monster. If man inflicted death as Nature has inflicted it he would be a monster. If man inflicted such painful diseases as Nature inflicts he would be a monster. Man rebels against these inflictions. Shall he add to pain by his rebellion ?. Is the infliction of pain by man, compared with the maximum of pain inflicted by Nature, justifiable?

It is a moot-point whether we ought to attempt to interfere with or lessen the pain of Nature by any method of research; or whether it is not our duty to accept pain as an act of Nature warning us of our errors, and telling us to remove those errors, rather than try to master pain by entering into experimental conflict with Nature on her own ground. Whether by any art of ours, beyond improvement in our knowledge of the ways of prevention, we embitter or embetter the manifestation of life, that is the question? Life might go on though our planet and everything on it were burnt to a cinder ; it is like time, nothing to us, and yet everything. Our business is with the bodily instrument, through which the living force does its work, and by which we recognise its existence ; our business is the study of matter moved by life —a study which brings us face to face with the phenomenon of motion we call pain.

Pain is the friction of matter set in motion by life. The scream of a sentient being in pain and the screech of an engine are sounds from

the same cause—friction of matter. Pain, therefore, tells its own story. It says, in the motion
of life there is, for some reason, undue friction.
A man is grinding mother-of-pearl ; he inhales
a hard dust in infinitely fine particles. The
particles enter his lungs, and without difficulty
pass through the vessels of the lungs into the
arterial column of blood that is worked by life.
They pass freely along with the blood into all
parts where the vessels are of sufficient size, and
where the surrounding parts are elastic they run
without friction ; but when they come to the
minute vessels of the cartilages they enter vessels
of smallest size surrounded by more unyield-
ing substance. The propelling motion tries to
force them on, but they cannot be forced ; then
comes the friction, then the excruciating pain
in the extremities, " pearl-grinders' neuralgia, or
rheumatism." Sometimes, from bad living, or
rather from improper living, we generate in
our own bodies similar insoluble particles of
matter, and, through the same stages of friction,
we get the same kind of acute articular pain.
We make pain in these instances unknowingly,
but knowingly by the rod, the wrench, the

ligature, the cautery, the knife; as a set-off we can annul pain by extracting blood, by applying cold, or by administering what are called narcotising agents—agents which modify the primary motion of material particles under the force of life.

What we wish to know is whether the necessary or unnecessary infliction of pain, or friction of parts of sentient beings, can be prevented; and whether in learning the art of prevention it is requisite to produce pain? It is curious knowledge to seek, and every honest physiologist must recognise the general interest it excites in the human mind, unless his inquisitiveness be more cutting than his wisdom, or unless his ambition runs ahead of his desire simply to be useful.

I have heard the foolish say that as pain is natural, so the infliction of it, artificially, is justifiable. But all pain is not natural. A perfectly constructed engine works with so little friction that it neither screams nor cries, and the same is the fact in a perfectly constructed living body. A perfectly constructed living body works without pain. The one natural pain is that of childbirth ; and if pain were not then felt, and aid called for, the highest forms of life could not hold a permanent existence. Children would be born while mothers were sleeping, and under varying conditions in which they could not live. Giving birth is an exceptional function, requiring exceptional labour, and causing, therefore, exceptional friction—pain ; but even this can now be artificially abolished if we will it.

If we look upon pain in its true meaning and service, it seems childish ever to inflict it for the purpose of preventing it. Granting that a perfectly acting healthy body can work from the beginning to the end of the chapter of life without feeling pain, except in such extreme emergencies that sounds or movements indicating the need of assistance are necessary, were it not better for man to use his powers to prevent friction or pain rather than to try to get at special means for alleviating it when it is permitted to present itself? The faithful answer to that question is, without doubt, in the affirmative. In this general effort all could work, gentle and simple alike, and, what is more, all engaged in such effort could work by simple methods. In the bright future all will work on one plan, and there will be no more pain. Our English people of to-day experience much less friction, and less pain, than those did who lived when Queen Victoria commenced to reign, and this—apart altogether from the art of relieving pain by artificial methods—is the result of scientific research.

But we are at present in a world of accidents, which lead to severe friction. Accident means displacement of working matter, matter worked by life ; it means pain. This is clear when we speak of a fractured limb ; but disease also, every disease, is an accident ; and, being beset with so many accidents, men are led often, against their inclinations, to accept risks they would not otherwise accept, in order to arrive at rapid solutions of protection and rescue. Some labour, in this direction, from real anxiety for the sufferings of humanity ; some for fame which, in this instance, is too often a glory on the water, which no one is more anxious, in the end, to see brought to nought than he who started it.

I have written this introduction to indicate

that I am in full sympathy with the belief that
the prevention of pain is one of the highest aims
of medical science.  But I thought it well, as an
experimentalist, to explain that the aims and
objects of experimentation on vital functions
are mainly promoted as a speedy and effective
method for removing disease and pain from a
world that is always making disease and pain
for itself ; is always wanting the animal engineers,
called Doctors, to cure both ; and, is forcing
some to experiment against time in order to
learn how to overcome what never ought to
exist in civilised communities.

I daresay some will expect me, in advocacy of
experiments, to use the common argument that
as man, in sport, and in slaughter of animals,
inflicts on the lower creatures all kinds of
tortures, therefore the animal engineers are
more than justified, even if they inflict torture
in their efforts after discovery.  I shall use no
such argument.  For the positive wants of man
the torture of animals is altogether unnecessary;
a remnant of an ancient barbarism, when man,
a savage animal, found it requisite to hunt
animals and to circumvent the cunning of

creatures of a lower nature, in order that he and his might subsist on what he captured. For the wants of man the slaughter of animals, for food, is equally unnecessary, a slur on Science that she cannot save man the expense and trouble of making the herbivora the laboratories of the carnivora, and cannot, by her own skill, produce all the necessities of sustenance out of the firstfruits of the earth. The comparison, therefore, of scientific experiments with barbarous methods is futile, because the methods are themselves corrupt, ignorant, and base. More than this, the comparison is ignoble. Men of science ought to be as much above the traffickers in sport or slaughter as goodness is above evil, learning above ignorance, wisdom above superstition. Scientists are neither vermin-catchers, sportsmen, nor butchers. They live to lift mankind into the liberty of health; and when they allow themselves to be compared with men of lower pursuits, they admit an affinity with them that brings them to a level they should shrink from if they were asked to assume it. I, for one, despise and disdain the argument. If we are warranted in giving pain to lower animals for our purposes of inquiry, we must use the warrant on another ground altogether—namely, that of absolute necessity—necessity for working out some great and beneficent object; and, we must use it in such a manner as shall show that, being really under the pressure of necessity, we are as anxious as the most sensitive to employ an exceptional power for an exceptional benefit, a benefit extending not to man alone but to all sentient existence.

Perhaps, it may be argued, that man, endowed with feeling equal to his wisdom and knowledge, might, exceptionally, exercise his authority for the infliction of pain on himself or on lower beings for the sake of sound research. I do not endorse all sentiments on this point, but I express my own distinctly, that we ought to have this exceptional right. How the exercise of that right may be rendered so exceptional as to be harmless will be considered in the succeeding chapters.

In the few sentences put together above I have, I trust, set forth fairly and honestly the belief that, except for useful purposes, I do not consider pain necessary nor disease necessary.

In a state of perfect civilisation, therefore, systematic experiments on sentient beings for scientific objects were an anachronism. In an imperfect civilisation like the present, when pain and disease exist everywhere, systematic experiments, even on sentient beings, may be exceptionally justifiable ; but the admission need not imply the necessity for the infliction of more pain, or a physiological *crux crucis*.

# CHAPPER I.

## INDISPENSABLE EXPERIMENT.

QUESTION 1.—"*In view of the difference of organisation between man and the lower animals, do you consider that painful experiment has played any indispensable part in the study of medical substances and methods for the cure of disease ?* "

THE word *indispensable* in the above question is the difficulty. Is any method of research in any direction of research indispensable ? I do not think it is. We are given to say, when we see that something done has led to something gained, that the something done was warrantably indispensable ! But, at best, this is only *post hoc et propter hoc* ; because if what has seemed to be indispensable had never been thought of, some other plan equally good would or might have led to the same results. The human mind is so inventive, so versatile, that nothing in method of learning can be considered as indispensable. The English language seems indispensable to the English people—seems so because it is a mode of expression springing out of many changes, and fashioned into a specific form, which, being convenient, remains for a time specifically our language. But that this form is not indispensable is shown by the fact that Englishmen can adapt themselves to other languages, and that our own language changes from age to age.

In Science there is no one method that can be considered indispensable. Attributes are indispensable ; observation, industry, accuracy, are indispensable ; methods are not. Methods may be convenient, they may be useful, they may be expedient, but nothing more. Methods run with the manners and customs of the ages. Celsus tells us that Erasistratus and the school he founded laid open the bodies of criminals in order to study by direct observation the action of the intestinal organs during existence—that is to say, in the state of life. The act at that date of civilisation, probably, shocked no one ; it was no doubt in accord with the spirit of the time. In a day not very remote from our own, a criminal, sentenced to death for some trivial crime, was given over to William Cheselden, surgeon to the first George of the Hanoverian dynasty, for experiment. The criminal was deaf, and the experiment intended was that of making a puncture through the drum of the ear in order to discover if an opening through the drum would enable the deaf to hear. At the last moment Cheselden, a man of fine feeling, and, as an operating surgeon, brilliant, declined the experiment, on which the criminal, whose life had been conditionally spared, was set free. For his failure, for his generosity of mind, for shrinking from an experiment on another human being, Cheselden lost caste at Court, and was considered pitiable by those who lived on courtly favours.

In our own age a physiologist, in his ardour to follow up the growth and development of the entozoa in the human body, administered to a criminal under sentence of death, at certain fixed periods of time before the execution, food containing cysticerci derived from the hog ; and after the criminal was slain he sought for the tape-worm, developed in its different stages, in the intestinal canal. The experiment was painless ; it was productive of discovery ; it was truthful ; it was one that Cheselden himself might have accepted ; but it was contrary to the spirit of this later time, and was, therefore, criticised. More recently many experiments have been conducted in America for slaying criminals by the electric shock. The object may, in a certain sense, have been considered humane ; but in its practical bearing it has created an absolute aversion towards some of those who have taken part in it ; and, has led, more than anything else, to the re-opening of the question whether capital punishment itself, in any form, is indispensable, with an increasing conviction that capital punishment is a barbarity that is not indispensable.

The argument is taking now the same directions against experiments by man conducted on the lower animals for the purposes of discovery ; and when, from the history of the past, we gather what has been achieved by such experiments, there is but one answer—namely, that such experiments, although they may achieve what was expected from them, were not indispensable.

They may have expedited discovery; they may have confirmed discovery; they may have led to discovery; but they were not indispensable. At all events, I, for one, who have no special reason to find fault with the methods, dare not assert their indispensability.

I observe in the question put before me, that difference of organisation between man and the lower animals is suggested as a reason why experiments on lower animals are not indispensable for the study of medical substances, and methods for the cure of disease. The difference must be acknowledged with the difficulty arising from it, and due weight ought to be attached to it. Dr. Weir Mitchell, of Philadelphia, observed that pigeons were practically unaffected by opium. He administered as many grains of opium to a pigeon as would have been sufficient to destroy the lives of two or three strong adult men not addicted to opium. I took up the same research, and found that pigeons could, for a time, live on opium as if it were a food, without showing any indication of narcotism, such as is seen in man, unused to the drug, when it is taken even in a small dose. I found that the goat would take tobacco in what would, really, be poisonous doses for man, and suffer no injury.

The comparison of pain as between a lower animal and man is also of little value. In the human species itself comparison between the pain of one person and of another is itself of little value. I presided once over a committee of inquiry, whose duty it was to report on different attempts to extract teeth without the infliction of pain. In this instance the same kind of operation, performed always by a skilled operator, was attended, in different persons, by symptoms so entirely contradictory that nothing satisfactory could be determined upon, or recorded with fidelity. Amongst the lower animals comparisons are even more difficult to make, and certainly no animal can in any way compare with man in respect of sensibility to pain. In the face of so much difference as to sensibility, experiments bearing on pain are open to the most serious error. I thought at one time, from observations made on the human subject and on lower animals, that I had discovered a simple and effective mode of producing

local insensibility to pain—local anæsthesia—
by which surgical operations could be rendered
painless. I called the process "voltaic narcotism," and under it, as it seemed to me, operations that would have called forth every
indication of suffering in a lower animal, like
the dog, if no anæsthesia had been induced,
were performed with absolute immunity from
suffering. But another physiologist, the late Dr.
Augustus Waller, repeated these same operations
without using, what I thought, the essential
benumbing process ; obtained the self-same
results, and denied the value of the process
altogether.

I have seen, since the time when those observations were carried out, so many more facts of
contradiction, or, at the best, of uncertainty, I am
bound to admit that pain is not to be measured
by experiment. Some operations by their very
vehemence kill pain; some that might be considered comparatively trifling cause extreme pain.
Some natures, both in men and lower animals,
are obtuse to acute pain ; others amongst men
are susceptible to the slightest infliction of it :
and this leads me to make another statement,

namely, that pain, when it is excited and sus-
tained in any animal, obscures and falsifies, for the
time, all the other vital phenomena which admit
of investigation.   Pain  up  to  a  certain  degree
quickens the circulation; beyond a certain degree
reduces  the  circulation.   The  respiration  varies
with  the  circulation;  the  temperature  varies;
the secretions vary, sometimes becoming profuse,
sometimes  being  checked  altogether;  all  the
functions  of  the  nervous  organisation  are  disturbed  by  pain,  and  the  will  is  perturbed,
becoming  either  violent  or  depressed  according
to the degree of intensity of suffering.   In plain
words, it is utterly impossible to observe natural
function under the shadow of pain either in man
or  animal;  for  he  who  tries  to  observe  under
such  circumstances must  make  so  many  allow-
ances  for  the  circumstances  under  which  he  is
observing it he finds it is extremely difficult, if
even it be possible, to be precise in his conclusions.

Some think that certain vital experiments can
only be useful when they are performed irrespective of the pain they involve; I, on the contrary,
am  certain  that vital  experiments,  to  have  any
value  at  all,  must  be  conducted  without  any

trace of the disturbing influence of suffering, whether man or lower animal be the subject of observation.

So far, then, from painful experiments being indispensable to successful experimentalising, I consider them perplexing and deceiving, and I am borne out in this view by the fact that every experiment I have ever witnessed, and every surgical operation I have ever witnessed, has always been most satisfactory in respect to results when the experiment or operation has been conducted under conditions in which the sensibility was destroyed. Nor do I stand alone in this view ; I have heard it expressed by the late Sir Benjamin Brodie, by Dr. Baly, by Sir John Forbes, by Dr. W. B. Carpenter, by Dr. John Snow, and by many others whom I have known.*

I think we might all be of one mind that for

---

* The personal susceptibility of men must, of course, be taken into account on this question. Thus Baly, in some respects the most accomplished and learned physiologist I ever met, was so sensitive on this matter of pain that I once saw him become faint when a deeply narcotised animal, under experiment, began to show the signs of movement indicating recovery from

the actual performance of research painful experiment is not only not indispensable, but often
actually mischievous.

This carries us a long way, but it does not
meet one grand difficulty. It does not settle
the question whether the infliction of pain and of
disease, avoided at the moment of experiment,
but following upon an operation of an experimental kind, is indispensable. In the study of
cure or of prevention of disease is it indispensable to produce friction or disease in an inferior
animal in order to trace out the cause of disease
or to effect a cure ?

It is not necessary in considering this point
to measure out the amount or degree of pain
induced in lower animals by the mode of research, since we may take it for granted that
no lower animal has ever suffered more than
man himself suffers under the diseases with

the narcotism. Sir Charles Bell shared in this view,
and his little-known, but great, predecessor in research
on the function of the spinal nerves, Alexander Walker,
would never undertake an experiment on a lower animal
for the purpose of vindicating his choicest theories.
Walker maintained that anatomy alone was all-
sufficient for his proofs.

which he is inflicted through exposure to what
are called natural causes of disease, causes he
does not as yet understand and is incompetent
to remove.   The question is one of pain natural
*versus* pain artificial.   It is the *crux crucis* in full
view, and I need not wander far in order to illus-
trate it.   There is a disease called cancer ; it is not
only the most inscrutable, but, at present, the
most firmly planted and, taking it altogether, the
most excruciatingly painful of diseases.   It has
also this speciality, that it afflicts lower animals
as it afflicts mankind, so that neither the beast of
the field, nor the fowl of the air, nor the fish of
the sea, is exempt from it.   Of the true origin
and cure of this disease we are nearly as ignorant
as we were in the ancient day when it was described as cancer because it seemed to lay hold
of the affected tissues with the clutch of a
crab.   In the presence of this disease the
difficulty arising from difference of susceptibility
ceases.   Many years ago one of my favourite
dogs died of cancer.   If I could have radically
cured that dog of the disease, I should not only
have saved it from suffering and saved to myself
an attached and faithful friend, but I should

have seen my way to cure, radically, a hundred at least of human friends who have since come under my care for treatment while suffering from the self-same fatal affection. I had no cure for the suffering and disease of that animal; and, under even our present knowledge, there is no known method of curing this disease in man, woman, or lower animal. In respect to it universal medicine is daily and hourly making experiment. Every surgical operation for cancer, every remedy applied to a cancerous part, is an experiment; every drug administered internally is an experiment; every example of it left to take its own course is an experiment. Experiment of this kind, legitimate in its nature, as all admit, never ends. No one soever finds fault with experiment on these lines. Such experiment is sometimes called experience instead of experiment, but the change of word does not cover the fact. It is experiment, after all, important experiment, though as yet useless. If by it we could cure, all else would be useless, but we fail to cure by our ordinary experimental devices. In this dilemma the experimental physiologist comes forward and says, " Let me investigate." He asks, "Is cancer a local or a general disease?" and, to get an answer, he transplants cancer from one susceptible lower animal to another of the same species, in order to learn : —(1) If the morbid growth will continue to grow in a new soil :—(2) If the morbid growth is capable of development under every condition of development, or if it demands from the body it affects a special constitution "to back it up," if I may use so familiar an expression :—(3) If by the infliction of remote injuries on nervous centres or trunks connected with glandular structures, he can institute changes which may lead to malignant developments in those structures as a result of the nervous derangements he has experimentally induced :—(4) Or again, if he be of those who look on bacteriological research as the one research required in medicine, he may perform experiments in order to ascertain whether the disease is communicable, like a parasitic disease, such as scabies, or Madura foot.

Let us assume finally that an experimentalist, avoiding all vivisection or suchlike experiment involving necessity for pain, went back in research to the breeding of cancer in a lower animal. Assume that, having a male and female dog each affected with cancerous growths of distinct types, he bred from them another dog, and found that cancer affected the offspring. Would this experimentalist be less objectionable than he who transplanted a cancerous growth from one susceptible animal to another? Most persons would answer the question in the affirmative, because a " natural " method was allowed to enter in by the breeding, whilst an artificial method was introduced by transplantation. But the point is a very nice one, ethically, since, if either inquirer produced cancer in the course of his experiment, he would inflict the suffering of cancer, in an equal degree, on an innocent and a helpless animal.

I believe there is not one reasoning person who has seen the purgatory of cancer who would not say that every experiment, hitherto performed for the prevention or cure of this disease, has been, when carried out with a true and honest intention, a justifiable experiment. Our emotions, not less than our reasoning faculties, declare this fact ; and we may be sure that so long as cancer is an incurable natural infliction, men in their present state of knowledge will experiment after cure. But whether the infliction of experiment, as it is commonly understood, is indispensable, is another and an open question.

It will be my task to reconsider this matter in the answers to Questions 4 and 5. Meantime, I would apply what has been said—concerning the investigation of the disease cancer by experiment—to every other disease equally painful and obscure and equally resistant of prevention or cure. Experiment may be expedient, it is not indispensable.

# CHAPTER II.

QUESTION 2.—*"Has painful experiment on animals played any indispensable part in the particular case of the discovery of anæsthesia ?"*

IN the discovery of anæsthesia, general and local, painful experiment on animals has played no indispensable part whatever.

This answer is unmistakeable. The lower animals have been permitted to share, more than equally with man, in the blessing of anæsthetic discovery, for by it many of them have been saved the agonies of painful death ; but they have been subjected to painful experiment in the course of discovery. It is very important to put this matter beyond dispute, because we observe often that enthusiastic advocates of experiment are apt to point, triumphantly, to anæsthesia as proof direct of the beneficent gains from experiment in the art of the subjugation of pain.

The introduction of anæsthesia was an empirical introduction. It sprang from the observation of the action of various vegetable substances on the bodies of men. We are in darkness respecting the first observers of the phenomena of narcotism produced by narcotics. We are in ignorance as to which narcotics were first employed, whether the juice of the poppy, of henbane, of belladonna, of hemlock, of mandragora, but we suppose that the earliest narcotics were juices derived from vegetable sources.

For the application of anæsthetic substances to surgical art we go back as far as Dioscorides, who tells us about mandragora, and gives a formula for making from it a wine to which, later on, the name of *morion*, or, as we may call it, *death wine*, was applied. The first anæsthetic potion was therefore swallowed like wine, and it seems to have been effective. The swallowing of the narcotic draught was followed by a long sleep, resembling the sleep of death, under which operations could be performed by the knife or the cautery without the infliction of pain ; but when the patient awakened out of the sleep he or she was affrighted as if by some terrible

dream, and woke up with screams and terror.
It would seem also that some persons partook of
mandragora by habit, and were termed "man-
dragorites," or, as we should say, mandrakes ;
and hence the vulgar error, conceived in a later
day, that the plant mandrake itself screamed
when it was torn from the earth.

So profound was the sleep and the insensibility
caused by wine of mandrake, that many services
besides the surgical have been attributed to
it.   There is a Jewish tradition that during the
Roman occupation of Palestine the Jewish
women, under the sanction of the grand San-
hedrin, would go out to the Roman victims of
crucifixion, and administer to them the death
wine on a sponge, whereby the victims were put
to sleep and their sufferings were abated.   The
poets, too, had their interpretation of the services
performed by mandragora, and Shakespeare,
in the play of *Romeo and Juliet*, revived the
tradition by making Juliet drink of so strong a
soporific that, being alive, she might present to
those around her the signs of death.   He has
explained to us, by the proceedings of Friar
Laurence, that the sleeping poison was of

vegetable origin ; and in the symptoms that are
described he has depicted with skilful, as well
as learned, touch the phenomena which would
be induced by mandragora, with the addition of
the last consoling and cheering line about the
awakening as from a pleasant dream.   The plant
mandragora is allied to our deadly nightshade,
" Atropa Belladonna."   It is the " Atropa Mandragora," a plant growing in the Grecian Archi-
pelago.   For many centuries it was used as a
narcotic, and then, as is the case with so many
methods and practices of medicine, it fell into
disuse, and remained a mystery.   Twenty-four
years ago, the late Mr. Daniel Hanbury, F.R.S.,
collected some specimens of mandragora whilst
he was travelling in Greece, and gave them to
me.   From them, after the prescription of Dioscorides, I made once again " morion," and tested
its effects.   The phenomena repeated themselves
with all faithfulness ; and there can be no doubt
that in the absence of our now more convenient
class of anæsthetics, morion might still be used
for general anæsthesia.

The first steps of discovery, therefore, leading
to general anæsthesia were not experimental in

the sense of experiments of a painful kind inflicted on lower animals. They were the ordinary
empirical experiences derived from man himself
on subjecting him to the action of medicinal substances, and tried in practice when, by frequent
repetitions of experiences, their virtues were
known. The late Dr. Snow, in his posthumous
treatise on chloroform and other anæsthetics,
very industriously studied this curious historical
subject, and pointed out, from the Chinese litera-
ture, that a preparation of hemp, called by the
Chinese " Ma-yo," was administered prior to
operation, and that the patient was by this means
soon rendered as insensible as if he had been
intoxicated or deprived of life, under which
condition openings, incisions, and amputations
could be carried out painlessly. This practice is
said to have been performed by Hoa-tho, a
physician who flourished in the dynasty of Wei,
about 230 of our era. There is nothing remarkable in this statement, for we know that
cannabina, the active principle of the Indian
hemp, when swallowed in sufficient quantity,
induces a special intoxication attended with
general insensibility. In the thirteenth century

3

of our era, Theodoric, an Italian author, describes
an attempt by Dominus Hugo to produce
general anæsthesia by inhalation of the fumes
rising from a mixture of vegetable juices,—henbane, hemlock, lettuce, mandragora, and some
others of less potency.  It is strange that, having
advanced so far, surgeons did not think more
about inhalation.  They did not, but went back,
so to say, to potions administered by the mouth;
and in the course of last century it was a
matter of controversy whether or not it were
safe to administer a narcotic such as opium to
the extent of destroying general sensibility.  On
this point Dr. Silvester republished in the old
*Medical Gazette* (vol. xli., p. 515) a curious record,
in which it is shown from the *Skizzen* of A. G.
Meissner, published at Carlsruhe in 1782, that
Augustus, King of Poland, having received a
wound of his foot which led to mortification,
Weiss, his surgeon, a pupil of the famous
Petit, induced sleep by a narcotic draught, and
during the sleep removed, with success, the
decomposing parts.  The operation roused
the patient, but on being soothed he went
to sleep again, and did not know until the

following morning that the operation had been performed.

These researches indicate an extended and imperfect second stage of discovery leading to general anæsthesia ; and it is worthy of remark, that the observations were all conducted on human beings ; they were empirical attempts based on a rational idea, and springing out of ordinary practical observation of the action of drugs on men. There was not an instance in which any lower animal was made the subject of experiment.

So much is true as to the original intention of general anæsthesia or that anæsthesia which, under the influence of some chemical sub-stance, by one act, produces insensibility to pain throughout the whole system. It may, however, be advanced that the modern method of per-forming this act is different from the ancient, and it may possibly be admitted that if the first observers had commenced their inquiries by ex-perimentation on the lower animals with opium and other active drugs, they would have been deceived by the results they obtained, and that the grandest remedy ever given by man to man, *opium*, would never have been brought into practical use for the benefit of suffering humanity.

But it may be argued that the modern methods of anæsthesia are far superior to the ancient; that they are conducted with more precision and with more perfection than in previous times; that, in a word, we have now organised a system which works like a mechanical device; and, that the anæsthetist governs pain, as the engineer governs the engine, by turning on or off his narcotising vapour at his pleasure.

This wonderful acquirement, perhaps the most wonderful of all human advances in medicine, has resulted, it may be urged, and is urged, from experimentation on lower animals. The old system was empirical and accidental; the present is experimental and evolutionary.

Is this the fact?

It is not the fact. The present system proceeded in its beginnings on precisely the same lines as the old system. It began out of empirical observation; the only point of difference between the two being that the modern system has been founded upon the empirical observation of the action of vaporous or gaseous substances that will enter the body by the lungs instead of substances that will enter by the alimentary canal.

The present method or system of general anæsthesia sprang out of the pneumatic chemistry originated by the illustrious Dr. Joseph Priestley. In this research Priestley discovered the first, and still most extensively used, of the gaseous anæsthetics. This was nitrous oxide gas, a substance which passed into the hands of Sir Humphry Davy, who, from a series of experiments performed on himself by inhalation of the gas, came to the conclusion that "*as nitrous oxide in its extensive operation appears capable of destroying physical pain, it may probably be used with advantage during surgical operations in which no great effusion of blood takes place.*"

This text, on which modern anæsthesia was founded, this text of Sir Humphry Davy, the discoverer of modern anæsthesia, was pronounced in the first year of the nineteenth century, and of all presages of the century, now so near to its end, was one of the most memorable. But even with so plain a text before the world discovery had to wait for forty years. Curious phenomena were observed from nitrous oxide.

Breathed, diluted with air, it excited those who
breathed it to intolerable laughter; philosophers
of the gravest sort split their sides with meaning-
less laughter when they were under its influence;
for it seemed to play the double part of exciting
the condition for laughter, and of suggesting
some unknown cause for the explosion.

During the courses of chemical lectures each
year, when nitrous oxide was the subject of the
lecture, learned chemical professors would give
a treat to their students by letting them in-
hale the laughing gas, and wild and furious
were the tricks that were played under its influence.  The late Professor Turner, of University
College, held quite a gala day on nitrous oxide
day; youths laughed till they cried, punched
each other without mercy, and sometimes, like
the priests of Baal, stabbed themselves with
lances, but did not feel.  Still discovery waited.
Surgeons did not see their way to the application of the signs before them.  They were not
chemists conversant with the manufacture and
manipulation of the gas; they were not physiologists, ready to appreciate what the illustrious
Davy had whispered to them; they were not so

sensitive to the sight and sound of pain as the rest of the world; indeed, they rather prided themselves that in respect to the proofs of pain they were not as other men, and so they went on, in the old way, little loath.

As is common and natural in the course of human discovery, curiosity prompted advancement. Observers began to strike comparisons, and one observer, great as Davy himself, struck a comparison as between the effects of the vapour of sulphuric ether and of nitrous oxide gas. This observer was Michael Faraday, the pupil, friend, and successor of Davy. Faraday, though not a physiologist, made the new observation that when *vapour of sulphuric ether* was admixed freely with air, and was inhaled, some of the peculiar symptoms inducible by the nitrous gas were presented. This was step two in the progress of discovery, and became, in due time, matter of demonstration in the chemical classes. Turner was prompt to show the comparison between the effects of ether and laughing gas in class. I knew, personally, a student of Turner's who, to use his own language, "got drunk twice in one morning in the class-room,

once from gas, once from ether," the gas being
the most pleasant but most evanescent in its
effects, a fact we understand well enough in
these days, and can explain on physical principles.
Still discovery lingered.  The administration
of laughing gas became a show, and surgery
went on bearing its burden of pain.  At last
from a mere show came the divine secret so
singularly concealed.  The practical lesson from
the text of Davy occurred at an exhibition
intended for the amusement of a public audience.
It was truly a strange passage in human history
when practical anæsthesia first saw the light,
when the practical art Davy had suggested burst
from its concealment, and in a moment, as it
were, was brought into practical working for the
removal of pain.  On December 11th, 1844, a
Mr. G. Q. Colston, a popular man of science,
appeared before an audience at Hartford, Con-
necticut, U.S.A., to lecture to, and to amuse
it, by experiments on chemistry and on
laughing gas.  Colston, I have heard, had been
a pupil of Professor Turner, and was well
acquainted with the most striking demonstrations
applied in the class-room ; he was also an able

and persuasive lecturer, and so fair a chemist, that if his mind had been as devoted to research as it was, by presumable necessity, to the making of a living by the presentation of the wonders of science for the delectation of the masses, he had possibly taken a first place in discovery. As it was, he merely gave a lecture in a popular and illustrative form.

In the audience gathered round Colston on this December 11th, 1844, was a dentist of Hartford by the name of Horace Wells. Wells, as we gather from what is left concerning him, was a nervous, emotional man, quick to perceive, rapid to act. At this moment, by a fortunate chance for the world at large, he had a troublesome painful tooth, and when the popular lecture was over, he invited Colston to his own office, or consulting room, and called in the aid of another dentist named Dr. Riggs. Wells requested Colston to administer the gas to him until he was unconscious, and asked Riggs, at that moment, to extract the aching tooth. Both acts were performed. The tooth was extracted without consciousness of pain, and when the patient came to himself he exclaimed, " A

new era in tooth-pulling." It was, in fact, a
new era in surgery ; it was the practical birth
of general anæsthesia.

It was not long before the practical discovery
was extended. Another American, named Dr.
W. T. G. Morton, who was acquainted with
the similarity of action between the vapour of sulphuric ether and of nitrous oxide gas, bethought
himself of supplanting the gas by ether. Morton was a man of greater ambition than Wells.
The act of Wells was, I feel sure, spontaneous ;
had nothing to do with the suggestion of Davy,
and was limited to the use of the gas in his professional task of tooth-pulling. Morton had a
wider aim. He saw that if sulphuric ether would
replace the gas, the means was at hand for
establishing anæsthesia all the world over as
a crowning mercy in surgery. It would appear
that he took into his confidence a professor of
chemistry, Dr. C. T. Jackson, who afterwards
claimed the entire credit of the advancement,
and for a time the two worked together. The
result was that, on September 30th, 1846, Morton, who was also a dentist, and who had once
been in partnership with Wells, narcotised a

man with the vapour of sulphuric ether, and extracted a tooth without causing pain.

The story of the further development of this process is most instructive, and belongs essentially to a book that deals with experimentation and the abolition of suffering. Morton and Jackson carried the method of inhalation into the operating theatre of the Massachusetts General Hospital, with the late Dr. Warren as the first operator ; speedily the news of the discovery arrived in this country, and became at once the event of the medical schools and medical societies in all the large centres. In London the first demonstration of the discovery was made in the house of Dr. Boot, of 24, Gower Street, where the operation, the extraction of a tooth, was performed by Mr. Robinson, a friend of Dr. Boot, a dentist in large practice in the same street. The patient was a lady named Miss Lonsdale, and the operation was painless.

Robinson, whom I, myself, afterwards knew extremely well from becoming connected with him in founding a practical school of scientific dentistry in England, has often related to me the details of this first important modification in the practice of English surgery.

The news came to Dr. Boot in a letter from America,* in which the details of the administration of ether were supplied. The letter was written by the late Dr. Bigelow, of Boston, and was peculiar in more ways than one. The discovery had been originally announced and demonstrated by Morton; but, unfortunately for his lasting fame as a discoverer, not in a clear, simple, and open exposition. He treated the agent he employed as a secret worthy of a patent, and called it "Letheon." It was an attempt at concealment as foolish as it was unworthy, for the odour of "Letheon" at once betrayed to Bigelow that the substance was nothing more than pure sulphuric ether. Bigelow, so soon as he detected this fact, commenced to administer ether with success, and communicated the circumstance promptly to all he knew, as well as to his English friend Boot. Boot, on his part, was not slow to call in the assistance of Robinson, in order to put the experiment to the test on this side the Atlantic.

Robinson told me that he had just breakfasted, on Dec. 17th, 1846, when he got the startling

* The letter was brought over by the *Arcadia*

intelligence from his neighbour about ether-inhalation. A man of great enthusiasm and of quick action, he was round at Boot's " in a jiffy," expressing himself ready and anxious to have " the first fling " with ether. A patient was soon found in the lady, whose name I have given— Miss Lonsdale—who was about to be operated on by Robinson under any circumstances, and was " proud to be the first taster of painless surgery in England." Robinson was a little troubled in making sure of pure ether, and when he was satisfied on that point he had to study how best to apply it as vapour. He devoted the 17th and 18th of December " in rigging up an apparatus," out of a " Nooth's inhaler," armed with a flexible tube and mouthpiece ; and on the morning of the 19th, in the presence of Dr. Boot and his family, at Boot's residence, he put Miss Lonsdale to sleep in a minute and a half, extracted a molar tooth from her lower jaw, and saw her restored to consciousness and safety within a minute afterwards. When Dr. Boot questioned her about the extraction, she expressed the greatest surprise at finding the tooth was removed. All she had felt was a sensation of coldness around the tooth, caused, Robinson thought, by the coldness of the extracting instrument.

On December 21st or 22nd, the famous surgeon, Liston, amputated a limb painlessly, under ether, at University College Hospital, the anæsthetic having been administered by Dr. William Squire, who is still in practice in London. Amongst those present on that occasion was my old friend, Dr. (afterwards Sir John) Forbes, the author of "A Physician's Holiday." He described to me that he never felt so near to falling on the floor in all his life as he did when he witnessed the great surgeon Liston amputating a thigh while the patient was in deep sleep. In those days, in order to save pain, the surgeon cultivated rapidity of action, and such an adept was Liston that he completed the removal of the limb within the minute. This, combined with the momentous result of the annihilation of pain, was the cause of the sensation experienced by Forbes. It was not fear, it was not faintness ; it was an emotion painful, as he expressed it, from its overwhelming surprise and pleasure. Everybody seemed pale and silent except Liston, who was flushed, and so breathless, that when he broke the silence with the word "Gentlemen," he almost choked in its utterance.

From these simple beginnings sprang up the grand art of anæsthesia. In a short time the value of the art was attested by the discovery of its power in preventing that special act of suffering under which the one half of the higher creation groaned. Dr. James Simpson, of Edinburgh, in the first month of the year 1847, administered ether, successfully, to a woman to allay or quench the pains of childbirth; and when, afterwards, he was charged by the strictly religious world, that he had infringed the natural law, "In sorrow shalt thou bring forth children," he answered triumphantly that the first recorded operation ever performed on man was that performed on Adam, when there was removed from him a rib out of which to make woman, and that before commencing the operation God Himself cast the man into a deep sleep.

The instauration of general anæsthesia came from experiments made on man alone. There is no suspicion of any experiment on a lower animal in connection with it.

But man is a restless and inquiring animal. He never gets a new thing or new art without trying to improve it, or, at all events, supplant it, by something he thinks to be newer and better ; so, naturally, and as a matter of course, he soon began to ask if he could improve on ether as an anæsthetic. Ether was slow in its action ; it had a disagreeable odour, and it required a rather complicated apparatus for its administration. These were objections which had to be removed, and many began to seek for a better substance with which to master pain. The word *ether* got the first place, and held the field by virtue of that priority. Whatever new anæsthetic was introduced it must be an *ether*, and the next thing that took the field was another ether, in name at least. Since the year 1831 there had been in use in medical practice a fluid called *chloric ether*. It differed from what was known as sulphuric ether in that it was a compound or mixture, two substances commingled together. Sulphuric ether, commonly known simply as ether, is a fluid substance consisting of three elements, carbon, hydrogen, and oxygen, and made by the distillation of alcohol with sul-

phuric acid. It has distinctly its own specific
qualities, its own specific weight, its own vapour
density, its own boiling point, its own rate of
diffusion, and its own solubility in other fluids,
like water, alcohol, or blood. Chloric ether
is not of this simplicity and unity of action.
It has two parts, each part possessing its own
particular properties. In the year 1831 an
American chemist, Mr. Guthrie, distilled together
chloride of lime and alcohol, by which he obtained an alcoholic solution of a substance, to
which this name chloric ether was applied by
another distinguished chemist, Professor Thomas
Thomson, of Glasgow. The alcoholic solution of
this volatile and pungent substance went under
this name of chloric ether, as it does to the
present day. But what was the substance con-
joined with the alcohol? Liebig separated it
and named it *chloride of carbon*, under the incorrect idea that it did not contain hydrogen. This
incorrect idea was put right by Dumas, who also
separated it from the alcohol, and by better
analysis, fully described and defined it, and gave
to it the name of *chloroform*. From that time,
chloric ether was made methodically by mixing

4

together chloroform and alcohol, and to the
initiated it was known to be a mixture of this
character, but to the world at large, and even to
the medical world, it passed and passes by its old
name, chloric ether, ranking as an ether. When,
therefore, improvements or variations began to
be considered in anæsthetic practice, it was
natural enough that some one, keeping in mind
the word " ether," should suggest the trial of
chloric ether. Two men quickly thought of it
in this sense, and gave it trial. The one was Dr.
Bigelow, of Boston, U.S.A., the other Mr. Jacob
Bell, the then well-known chemist of Oxford
Street, London. Dr. Bigelow did not succeed
well in his trials, and gave the inquiry up for
the moment. Mr. Bell was more successful; he
caused anæsthesia in a considerable number of
instances by means of chloric ether, and was
struck, he told me, by the perfection of the sleep
it induced, although perplexed and dissatisfied
by the slowness of its action. He was able,
however, to get the new ether tried for operations in the Middlesex and Bartholomew's
Hospitals. At Bartholomew's Mr. Lawrence
operated on patients under chloric ether, and

was so satisfied with the results that he applied the anæsthetic in his private practice, thinking it was an ether.

It is the peculiar part of the course of discovery at this stage that what was effected by chloric ether was due, not to an ether at all, but to the substance which Dumas had called chloroform, and which was present in the mixture in the proportion of about twelve parts by volume of chloroform, to eighty-eight parts of alcohol. It was not surprising that Mr. Lawrence, an operating surgeon of the strictest sort, should not appreciate this fact from a chemical point of view ; but that Mr. Jacob Bell, a man whose life was practically devoted to the study of chemical preparations, should not have detected the active principle he was employing, is one of those curious circumstances in research that baffles calculation. In fact, Bell never could account for it himself ; and I have heard him say that whilst he knew perfectly well that the so-called chloric ether was a mixture of chloroform and alcohol, —he having made the mixture with his own hands—and whilst he would have known in an instant if the thought had occurred to him that

the alcohol could not possibly cause the narcotic
action, and that, consequently, the chloroform
did cause it, the fact never occurred to him to
solve the riddle, being misled, as he believed,
by the use of the word ether, and by the feeling
that it was an ether, and nothing else, that was
demanded. Had he by the merest chance been
moved to the explanation of the effects that
were manifested, he would undoubtedly have
been the acknowledged, as he was the un-
acknowledged, discoverer of chloroform as an
anæsthetic.

Chloroform, then, was in use in practice before
it was known to be ; but, as a matter of necessity, the secret soon came out. How it came
out may rapidly be narrated. Mr. David Wal-
die, of the Apothecaries' Company, Liverpool,
an excellent pharmacist, who had been making
chloric ether by adding chloroform to pure alcohol, divined what Bell did not divine—namely,
that the substance in chloric ether which caused
the narcotic sleep was the chloroform it contained. Being in Edinburgh in October 1847,
he told the fact of the use of chloric ether in
London to Dr. Simpson, explaining that the

substance at work was chloroform, and, recom-
mending Simpson to try chloroform, promised to
make him a specimen for such trial. Some little
delay took place in the fulfilment of this promise;
and meanwhile Simpson, having got chloroform
manufactured for him in Edinburgh; experimented with it in its pure state; and on
Nov. 10th immediately following—1847—read a
paper on the subject to the Medico-Chirurgical
Society of Edinburgh; instantly published the
paper under the title, "Notice of a New Anæsthetic Agent as a Substitute for Sulphuric
Ether," and began to use chloroform for the ex-
tinction of pain in the human subject. Very few
scientific medical essays have ever attracted so
much attention as this on chloroform. Chloroform soon came into almost general use in place
of ether, and the word itself became so common,
in the vernacular, that the people began to recognise it as synonymous with, and more expres-
sive than, anæsthesia. The word entered into
different parts of speech; those who administered the liquid were called "chloroformists," and
patients were said to be "chloroformed," as if the
word "chloroform" made part of a new verb.

The story of the introduction of general anæsthesia as an instauration is now told. I repeat
that the discovery it relates has in it no evidence
of advancement by and through painless experiments on the lower animals. Man all through
was the subject experimented on, in which process
he accepted risks but no real pain. There is a
statement afloat that rabbits were put to sleep
under chloroform by Simpson before he and his
friends inhaled the vapour ; but it is certain that
chloroform had previously been administered to
the human subject, and it is equally certain that
any animal subjected to the vapour was merely
put painlessly to sleep, or to death, by sleep,
under its action. These experiments, therefore,
painless in themselves and harmless to the
lower animals, did not lead to the discovery nor
aid it. On the contrary, there is a most notable
fact in relation to experiments under chloroform,
made on lower animals, which suggests that if
they had been relied on chloroform would never
have been introduced into practice. Flourens,
the eminent French physiologist, tried the effect
of chloroform on inferior animals, and, in con-
sequence of its powerful and fatal influence on

them, put it aside as an anæsthetic, rightly, as they who are opposed to it now declare; but, whether rightly or wrongly does not affect the present question, in so far as it relates to the discovery and the introduction of anæsthesia, one of the greatest blessings science ever gave to man.

## LOCAL ANÆSTHESIA.

Up to this time I have treated only on general anæsthesia, but there are methods for producing local insensibility to pain which have been tried, and which deserve notice in this place. These methods are : (1) By the application of benumbing pressure to nerves. (2) By exhaustion of blood from a part. (3) By localised electrical shock. (4) By voltaic narcotism. (5) By the local application of narcotics. (6) By the application of benumbing cold. (7) By means of a painless cutting instrument.

The effects of benumbing pressure were first tried by Mr. James Moore, son of the well-known author, Dr. James Moore, and brother of the famous hero of Corunna, Sir John Moore. James Moore thought that by bringing firm pressure

to bear on the trunk of a large nerve he could
cause anæsthesia in the parts it supplied, and he
invented an ingenious compress for this purpose,
depicted in my *Asclepiad*, vol. vii., facing p. 355.
The process, tried in St. George's Hospital, was
not found sufficiently practical. It caused numbness, but it was painful in its application, and
was very slow in its action. The experiments
with it were all conducted on the human subject.

In 1862 I made an attempt to carry out local
anæsthesia by exhaustion of blood from a part.
I noticed that when three round cupping glasses
were applied to the body very close to each
other, the clear triangular space left free within
the rim of the mouths of the glasses was
rendered white, brawny-like, and insensible,
when the suction of the glasses was complete.
This was obviously due to the local abstraction of
blood from the part, and I thought, consequently,
that if I could exhaust the blood from the ex-
tremity of a limb, the exhausted part might be
operated on without pain. I invented, therefore,
a boot, like what was once called Junot's cupping
or exhausting boot, but with this difference, that
the foot could project out of it, through a broad

indiarubber ring, and could be exhausted of its blood by extracting the air in the boot by a small air-pump. I tried the process on myself, and finding it succeed, the operation of removing the nail of a great toe was carried out under it on a patient quite painlessly, the patient looking on and feeling nothing. But the proceeding was too long and cumbersome to admit of introduction into practice generally, although it indicated an important principle, which may in some future day be utilised. In this research no experiment on a lower animal was resorted to. I was myself the victim in all the preliminary experiments.

• In the year 1859 attempts were made to apply electricity for the local abolition of pain. Rapid intermittent currents were passed through portions of the body, one pole of a battery being applied to the body, the other to the knife or forceps used by the operator. ˙ The idea was that the parts included in the circuit could be benumbed by the current. The plan was tried mainly for the extraction of teeth, and I presided over a scientific committee, by whom the method had fair trial. It seemed to me that

whenever the current acted favourably it acted in
the same way as a sharp blow on the head with
the flat hand diverted the pain like a sharp
pull in extracting a hair, an old schoolboy trick.
In fact, it failed, as the late Professor John
Marshall discovered, in cutting operations. Yet
it is only fair to say that in an experiment
Dr. Julius Althaus performed on myself, of
passing a very rapid current down my arm in
the direction of the ulnar nerve, a most distinct
numbness was produced, which lasted sufficiently
for a long operation had it been called for. In
these researches no lower animal was subjected
either to painful or painless experiment.

Soon after these experiences I introduced a
new method, which I named *voltaic narcotism*,
often referred to since by that term. It
consisted in applying over the surface of the
body, in small sections, a fluid, like chloroform,
or tincture of aconite, or a mixture of such
fluids, and then passing a continuous electrical
current through the part. Over the surface
covered with the fluid, on sponge or lint, a plate
of platinum or copper was laid, and connected
with the positive pole of the battery, the nega-

tive pole being applied, by a plate moistened with water, to some neighbouring portion of the cutaneous surface. The idea was that the narcotic could be carried or diffused into the structure beneath, and would then and there create sufficient local insensibility to enable the operator to cut the skin without pain. The results were most curious. Several minor operations, and one or two major, were carried out painlessly under voltaic narcotism ; but many objections came in the way of the adoption of the method. It was troublesome ; it required too much time, and it left, in some instances, an irritation which was not favourable to healing. I therefore gave it up ; but, singularly enough, it has within these last few years been re-introduced in America by Dr. Peterson and others under the name of " anodal diffusion." *

In the research on voltaic narcotism all the first experiments that led to its practical trial were on the human subject—all indeed that were necessary. After a time lower animals were subjected to it for operation, and with results which seemed to confirm its efficiency. These

* See *Asclepiad*, vol. viii., pp. 267-86.

observations upon the lower animals, the late Dr. Augustus Waller, of Birmingham, submitted to experimental criticism. He performed on similar animals the same operations without the aid of any electrical current, and obtained what appeared to him to be equally painless results. He held that I had been deceived, and that the immunity from pain was due simply and solely to the absorption of the narcotic fluids employed.

The simple direct action of narcotic substances for the production of anæsthesia was carried out, after Waller's essay, with considerable industry. It had already been attempted by the late Mr. Nunnerley, of Leeds, who deserves the first place in modern research on this matter. I carried out a long series of experiments with aconite, morphine, carbolic acid, and compounds of these, with varying success; and recently the most promising advance in this direction has been effected with cocaine, an admirable local anæsthetic, with perhaps fewer disadvantages than any other of its kind, and with the special advantage of practical readiness, and promise of better things to come.

Towards local anæsthesia no experiment of
either a painful or painless kind was made on
the lower animals that proved of any remarkable value. I do not think mine afforded any
service beyond showing that what was effective
on man was still more effective in animals of
lower sensibility.

The most numerous and extensive efforts for
local anæsthesia have been those in which
extreme cold has been employed to produce the
benumbing effect. The earliest applications of
cold originated between two and three hundred
years ago in the fencing schools of Naples. A
Neapolitan professor of training placed crushed
ice in a flask of thin glass, and then applied the
chilled glass to the skin, and held it there until
the skin was frozen, in order that the cautery
could be employed, or other small operations
performed without the infliction of pain. The
proceeding must have been most successful,
and why it became lost is one of the mysteries
of scientific research. It did remain lost until
our own time, when it was brought forward,
quite independently, by a man of true genius,
the late Dr. James Arnott. Dr. Arnott used

a freezing mixture of ice and salt for local
anæsthesia. A little later I invented for the
same purpose the ether spray process, in which
the benumbing cold was produced by projecting
a volatile liquid, like ether or amylene, or a
stream of compressed gas, like methyl chloride,
or carbonic acid, on the part to be anæsthetised.
These methods have been so widely adopted I
need not enter into any description of them.
I have merely to say that they were made
without any aid from experiments of a painful
kind on the lower animals. Dr. Arnott proceeded from the first on the human subject,
and I did the same. The earliest experiment
with ether spray was made on my own arm ;
the next was on a patient for the extraction of
teeth ; the next on a youth who required to
have a large portion of bone removed from
a lower limb ; and so the use of the spray
progressed up to such major operations as
ovariotomy and amputation of the breast.
Afterwards the method was most successfully
adopted for the purpose of making operations
on lower animals painless. I brought it into
use for the nerving of horses and various other

operations, for which use a foreign learned society honoured me by the presentation of its medal; but this was the result of an experimental service passing from man to the inferior creature.

The last research in local anæsthesia relates to the attempt to produce a painless cutting instrument. It was long known, before the days of anæsthesia, that the surgeon who could make a skilful incision with the greatest celerity was the most painless operator. Thus one of the patients of the famous surgeon Cheselden expresses himself in a poem addressed to the surgeon after an operation :—

> "So quick thy hand, 'twas hard to feel
> The progress of the cutting steel."

It has also been observed that in some accidents from cutting instruments, like circular saws and blades worked at rapid revolutions, and from projectiles from firearms, the injury has been so swift that limbs and portions of the face or trunk of the body have been removed without consciousness. These facts led me to invent the painless cutting knife, a knife that divides the flesh so rapidly there is no sensation

evoked by the act. The vibrations are so
intense that sensation is not realised. The
instrument was most successful in its working,
but no mechanic has yet produced it in such
a convenient form as to render it available in
practice. The experiments with the painless
cutting knife were all made originally on my
own body, and afterwards for operative purposes
—opening of abscesses and incisions of a super-
ficial kind—on other persons. A few operations
on lower animals were corroborative tests ; but
in not one was there any sign of pain from
the incision, and not one was indispensable
for discovery.

I have now put together, in as short a space
as possible, the course of discovery for the development of anæsthesia. It is fortunate for
me that I have been an eyewitness of the
progress made in this department from its
practical instauration. I recall the days when
operations were performed without the aid either
of general or local methods for abolishing pain.
I have myself introduced new methods of anæs-
thetising, generally and locally ; I have brought
to trial a large number of new anæsthetics. I

have utilised anæsthesia for the service of the lower animals as well as of men, and by the invention of the lethal chamber have had the delightful privilege of removing the taste and pain of death from probably a million of those friends of man, the faithful dogs. I write this not boastfully but truthfully. I write it in order to show that I venture to speak with authority in repeating, as an answer to the question submitted to me, that painful experiments on the lower animals have played no indispensable part in the particular case of the discovery of anæsthesia.

Whilst giving due answer to this important question, I must not hesitate to express that some experiments on the lower animals have been, if not indispensable, very expedient in the field of anæsthetic inquiry. These have not been painful experiments, and therefore may not be considered as coming strictly under consideration, but they have been made, and ought not, consequently, to be passed over as if it were prudent to omit them. I will set

5

them forth briefly from my own work, which will also represent fairly the work of other investigators.

There has been much search made in order to ascertain whether a safer anæsthetic can be found than chloroform, an anæsthetic that shall combine all the ready advantages of chloroform with fewer of its dangers. No anæsthetic has yet been discovered that is free of danger, but some, less convenient than chloroform—methylene, nitrous oxide, and ether, for example—are less dangerous.

The chemists are constantly finding, during their labours, new chemical bodies like chloroform, and we have been following them sharply to see if they can give us a safe anæsthetic, I mean a new fluid that shall safely anæsthetise. I have investigated as many as twenty-nine of these substances, and have, by this means, arrived at such a point of knowledge respecting them, that if I know the elementary composition of a substance, its specific weight, its boiling point, its vapour density, and its solubility in the blood, I am able, on theoretical calculation, to determine, without a single experiment, whether it will, on inhalation, produce general anæsthesia. More than this, I can determine how long it will take for the substance, gas or vapour, to produce anæsthesia, what the symptoms induced by it will be, and what time will be required for the man or lower animal under its influence to recover from its effects when it has been withdrawn. These points are deducible from calculations based on the constitution of the substance itself. They are most useful as far as they go ; but, unfortunately do not answer every question. There are vital points which they fail to reveal. They will not tell us whether, when a man or other animal is under the narcotic influence, the respiration or the heart will cease primarily if the inhalation be carried a trifle too far. Now here is a matter of extremest importance. If, in deep anæsthesia, the heart continues to beat while the respiration stops, the danger is great, but not necessarily fatal, because we can re-establish respiration artificially ; but if, in the same state, the heart stops while the respiration goes on, then the danger is fatal. Some anæsthetics stop the respiration first, and ether may be accepted as typical of this class.

Others are uncertain in the matter; chloroform
may be accepted as typical of that class.
Others stop the heart first. What is true in
this direction in respect to man is true in respect
to the lower animal. According to my observation the rule is the same all round.

How shall the rule be tested in the first instance; on a human being or on a lower animal?
Shall we try a first administration of a new and
promising anæsthetic on an animal that might
the same day be killed for food, or on one of our
fellow-men or women? There cannot be two
answers to that question. I have followed the
plan of observing the mode of death on lower
animals always before venturing to proceed
further with the use of a new anæsthetic; then
if all has appeared safe, I have inhaled the
substance myself; and, so assured, I have dared
to administer it for operation on the human
subject. I pursued that course in introducing
methylene bichloride, methylic ether, and methyl
chloride, and I should repeat it if another likely
compound came into my hands for investigation.
I do not enforce that the experiment is indispensable; but it is so expedient, that they who

are opposed to experimentation call for it when
responsibility waits. Sir William Fergusson, the
late great surgeon, told a Royal Commission he
did not remember any case in which vital experiments had been practically useful. In saying
that he forgot his own teaching. Dr. Snow
introduced a new anæsthetic called amylene,
but did not, as he afterwards confessed, make
a sufficient number of experiments with his
new agent on the lower animals before administering it to the higher animal, man. Had
he done so, he would have found that amylene, owing to its insolubility, separates in the
blood as a gas, and so produces the danger of
formation of bubbles of gas in the blood which
will not pass through the minute circulation
when the power of the heart has been much
reduced. Snow had a death from amylene in
a patient upon whom Sir William was about to
operate, and soon afterwards had another death in
a patient on whom Sir W. Bowman was about to
operate ; whereupon, as Snow told me, Fergusson
was severe, declaring that he, Snow, had not
made a sufficient number of preliminary experiments on lower animals. Some years later,

when I brought out methylene, Fergusson gave
me serious warning. " Have you tried your new
anæsthetic thoroughly on the lower animals,
or are you going to fall into Snow's error ? "
" I have submitted forty animals to it," I
replied. " It is not enough," he responded ;
"a solemn introduction like that requires a
hecatomb of animals."

Without going so far as Sir William Fergusson,
I must hold with him, that in the case named free
experiment is at least expedient—so expedient,
that if a discoverer of a new anæsthetic were
to bring his discovery into practice on man,
and were at once to meet with a death, in
so doing he would, probably, at the inquest
come under the hand of the law for rashness
or over-confidence in theoretical calculations—
themselves the result of experiment—or for
too sentimental a respect for lower animal
life.

There is yet one other research in which the
expediency of experiment on the lower animals
is prominent. Under any form of general
anæsthesia, what might be fatal asphyxia, what
might be fatal syncope, might occur. If it does,

what are the best means for restoration? Up to
this time we are weak in our knowledge on the
question suggested. We may, by some of the
rules now in vogue, be actually doing the very
thing we ought not to do, and be leaving undone
the very thing we ought to do. How shall we
learn means of recovery except on the lower
animals? We dare not experiment on men and
women even in the presence of the fatal accident. I confess I see here no means of advancement except by experiment. I do not say it is
indispensable, because some may urge that it is
not indispensable to use a remedy that invokes
the risk, or to master the risk when it has been
invoked, but in our present state of practice the
expediency is unmistakeable.

One word more before leaving the question
on Anæsthetics and Anæsthesia. I have explained that the anæsthetic process has been
turned to good account in putting inferior
animals to death without suffering. Out of the
human family the pang of death may be absolutely abolished over all the earth as the ages
run their course. This relief from the sting of
death is not indispensable—for all through the

ages past the animal creation has died without it, and the greater part die without it still—but surely, in its development, it is expedient, and, in every advancement of it, lethal death must first be, as it has been, experimental.

# CHAPTER III.

## LINES OF PAINLESS RESEARCH.

QUESTION 3.—" *In the study of human functions can you suggest promising lines of research without resort to painful experimentation ?* "

IN answering this question I am embarrassed by the richness of the field of promise that lies before me. It may be admitted at once, that whether painful experimentation be useful or useless, it has had one indifferent effect ; it has diverted the minds of men too strongly from methods of research that not only lie open to the curious mind, but which lie temptingly open. The human mind is so constituted it will never cease to investigate the ways and mysteries of life. That is perfectly certain ; but the human mind, even in this matter, is perverse also. It gets into a beaten track, and on that track it steers its way, however narrow the path may be. One mind jostles there with other minds ; forces its way against other minds ;

pushes forward, falls back, and begins and ends
its career. Meantime it fails to discern all paths,
save that it is on, and so while one path towards
successful discovery is crowded, other paths of
bright promise are left untenanted.

In the past fifty years physiology has become
so narrowed in its progress, so bent on making
discovery by vital experiment, and that alone, the
world has ceased to think of its work by any
other name than vivisection, and of its professors
by any other name than vivisectors. The world
has been unjust in this respect, for experimentation means infinitely more than cutting into the
bodies of living animals, and many experimental
physiologists have done good experimental work
who never took a scalpel into their hands, or
any other instrument for infliction of pain.
These, nevertheless, have been mixed up indis-
criminately with vivisectors. Thus I have heard
the late Professor Sharpey compared with
Schiff, and the humane Charles Bell with
Magendie. The explanation is not far to seek.
The men who are most resolute in experiment
most daring—shall I say most demonstrative?—
in the way of experiment, are men of marked

character. They dare to do, to act, to perform. They are the verbs of science. They say with Renan, *dire n'est rien ; faire est tout.* They themselves, strange to say, are sentimentalists, as keen as the opponents whom they despise. They are led on by experiment to experiment, hoping to find, without any reasoning beyond their findings ; and so they hold the eye of the world on their work for good or for evil. They are watched ; they are wondered at ; they are never beloved ; they are never even admired in the strict sense of true admiration ; they are sometimes hated ; they make no loved name ; they pursue an object, which, at its best, has in it least of conquest and least of satisfaction ; but they stamp physiology with their own seal, they overshadow philosophical physiology altogether, and they ignore the man who, reasoning from simple open principles, lets nature alone remain as his experimentalist, and draws his conclusions only from her works of beneficence or of torture.

In this chapter I am anxious to correct what appears to me the error of too exclusive a system. I am desirous to explain, what is not by any means new, but what to the world generally, and to the medical world largely, is truly a new physiology with nature as the experimentalist, and man as the observer and chronicler. There is not an experiment of a physiological kind, however small, however great, however painful, however free from pain, however methodical, however apparently accidental, that is not to be found in nature, would men industriously seek for it ; and in this labour of industrious search lies the field of labour that now requires to be followed. If it once became a favourite sphere of labour to let nature alone inflict the pain, and man make the cure, experimental pursuit by man would dwindle into nothingness, would probably become contemptible, and would certainly fall into a second rank, as something to be carried on, by necessity, solving some difficult observation, or proving some uncertain proposition, in the purest sense of an exceptional task. Experiment, in a word, would find its true place. In human work it would modestly follow, not ambitiously strive to run before, natural experiment. Subordinated to its proper position, experiment of a vital kind might lose some of its attractions in the eyes of the sentimental experimenters, who rest on experiment alone, and give reason the go-by ; but it would cease to give cause of offence to vehement sentimentalists of the opposite school, and would satisfy the more reasoning and temperate world.

I have said that in the method of research here broached there is the widest scope for work. It is true ; and were I framing a treatise specially dedicated to the exposition of this method, it would be difficult to know when and where to bring it to an end. I can offer in this place but a few illustrations.

We have been exceedingly curious, as physiologists, to ascertain the functions of the brain and nervous system. The curiosity dates from the earliest times of medical science. Hippocrates, with marvellous shrewdness, says the brain is the *metropolis of humidity* ; the great water-holder he means, as such it really is— water condensed and locked up as in a cistern— the great centre in which external vibrations make their impressions by the senses. Men have wanted to know what are the parts of this

wonderful metropolis of humidity ; what are its
passages, its centres of space, its centres of
activity, its inlets, its outlets, its trade, its com-
merce. They have looked at it as if really
it were a metropolis, and have named its
structures and parts as they might crudely name
those of a city. They have its major and minor
divisions, its highways, its bridges, its canals, its
islands, its roofs, its tints, its circumferences, its
centres. It is almost laughable to observe and
read how men of science have cut up this
metropolis into their own human conceptions,
founded on their own human designs. They
have accomplished all by the study of anatomy,
the study of dead matter, against which study
no one objects, but of which all approve, as they
do of the exploration of the regions and cha-
racters of the planet itself. But there came a
time when anatomy was not sufficiently satisfactory to the mind. The inquisitive wanted to
know more of the dead city. Vesalius and his
contemporaries revealed it, and, metaphorically
speaking, traversed and described it, when it lay
dead before them. But what about the city
when it is alive ? The great divisions, the hemi-

spheres, what do they do? The bridges, what do they do? The cavities, with their contained water, what offices have they? and so on through all the mysterious system. The curiosity was most natural, and at first the method employed for discovery did not seem unnatural. At all events, it appeared to be a ready method. It was like the ready method of building a tower whose top should reach into heaven. They said, Let us take this metropolis in its living condition to pieces, and then notice how, under the crush, it will perform its work and play its part, in relation to the body it directs and animates. They re-moved, therefore, or "ablated," portions of the living brain of different animals ; they divided the courses of its nerves ; they drew off the fluid from its ventricles ; they removed the organ, re-moved it in halves, removed the smaller brain from the larger, and the reverse ; they interfered with the great outway, the medulla, and in various modes, which I need not repeat, they endeavoured to arrive at function by perversion of action. They have left us still in much doubt ; they have discovered some things, but their leading discovery, I fear, is contradiction. I will not stay to blame the method—I could not ; for in a painless way I have taken part in it—that is to say, I have anæsthetised parts of the vitalised brain to discover what modification of function then occurs ; but I pass to explain that this is not a field of research that need be positively followed. There is another and less objection- able mode open, even in this abstruse subject of investigation. Nature experiments with the nervous system in her construction of it. She builds in every important animal her metropolis of humidity, and, according to the build, she supplies a different animal—an animal with different functions.

Were I a young student, ambitious to learn the functions of the nervous system in their vital condition, I would avail myself of this bit of knowledge of Nature as a constructor and of Nature as an experimentalist. We have at our command hundreds of differently constructed brains, and an equal number of animals, to which each one of the differently formed brains ministers differently. Let us proceed, then, to compare difference of function with difference of structure. Why ablate a part which Nature herself has

never formed, in order simply to see how an
animal that owns the part would go on without it?
I will undertake to say that the conscious act of
every animal is stereotyped in the anatomical
character of its brain, and that the time might
come when, during an examination, certain
functions being fairly described, the question
might be put:

"State anatomically the parts that make up
the necessary central nervous organisations for
the performance of stated functions, and show
how they differ from other organisations."

I conceive that the most splendid field of
investigation is here offered to the physiological
biologist. Every faculty of mind, every phenomenon of motion, can be explained by uniting,
with the observation of functions during the life
of an animal, the observation of the construction
of the nervous system of the animal. All consciousness, all motion, have but one nervous
origin ; but both are moved according to construction. Variety of form leads to variety of
function ; and in studying the one variety by
the other, the student is, as it were, learning
for himself so many modified instruments, and

6

noting the results ; in other words, he is allowing
Nature to do the experimental work for him
much more neatly than he could do it by the
most dexterous experiment, and is watching the
effect.   It will be said, probably, that this mode
of research is too mighty a task for students to
undertake.   It would necessitate that the whole
language of living function should be acquired ;
it would mean that every part of anatomy, comparative as well as human, should be learned
and expanded.   Is that a true objection ?.   I
answer that it is the strongest argument in
favour of the proceeding.   The want of knowledge of comparative anatomy and function is
the one weak of weakest points in physiology.
Every physiologist who questions himself, honestly, knows this fact, knows that there never will
be a true physiology until a true biology—true
by and through all its manifestations—has been
established.   In the very vastness of the labour
lies its worth.   It gives opportunity for discoveries
unbounded, and all on a line that will throw a
charm instead of a chill over the world at large.

   This one subject, of itself, is sufficient for a new
physiology that would occupy the attention of

all students for many years ; and as it presents itself to the mind it suggests, and opens, new lines in so many directions, there appears no end to them. It is a new mathematical course acquired from Nature.

Let us turn to one or two other very simple indications of research in these directions. Why do some animals hibernate, and some, similar, fail to hibernate ? Why does the dormouse sleep, whilst the common mouse remains comparatively active in cold seasons ? The secret must lie in the difference of organisation of the two different animals. What is the difference ? In all probability it is some anatomical variation in the nervous systems. The student who discovers this variation treads on the borders of another field. How is sleep modified by nervous organisation ? Is sleep an essential necessity for every active animal ? Do some animals require less sleep than others, and does the requirement run with the size and weight of the nervous centres ? Further, in animals that require long sleep—man, for example—how does the character of the nervous organism, the method of feeding it, the method of working it, modify requirement ?

Thousands, nay hundreds of thousands of observations, lie before the student who would institute these inquiries, without the performance of a single painful experiment by his own act and deed, Nature having furnished him with any number of her own acts, if he will but trust her and reckon up her behaviour under the various conditions of her own daily labours.

The influence of conditions surrounding the living body is another subject that has been most imperfectly observed, and in regard to which nature is rich to the last degree in experimental fact. It is astounding how, in investigation, the effects of surroundings are overlooked, simply because the effects are not observed and seized upon as experiments of Nature that will have their own way and ride over everything.

Experiments in one part of the world are not the same in result as they are in another part. It has been observed that in tropical climates the experimental trials with anæsthetics on warm-blooded animals are not identical, as to consequences, with the same kind of experiments conducted in cold climates. How could they be ? My late friend, Dr. John Davy, observed

that whenever in tropical heat he drew blood
from the vein of a patient, the venous blood
was of the same red colour as the arterial. In
tropical heat the course of the blood is so
quickened, and the external supply of heat to the
body is so free, that there is less need of combustion of blood, less combustion, and less product of
combustion. Nature, in other words, is pursuing
in the tropics a different experiment from that
which she performs in colder regions. She is
conserving power ; she is showing that in an
elevated temperature life can go on without so
active a consumption of material for burning as
is demanded in a lower temperature. Therefore
the red arterial stream of blood runs over into
the veins, and a comparatively low animal fire
animates the body. In this condition the action
of a narcotic vapour must of necessity be very
different to what it would be when in a colder
climate it enters the blood in a more condensed
form, and charges blood, that is consuming and
giving out its vital heat, with comparative
difficulty. It requires no new experiments to
account for any difficulties evidenced under such
distinct conditions, Nature being so powerful

a controlling influence as to vitiate artificial experiments altogether, if her own experimentation be left out of the record.

The same reasoning would apply equally to the influence of Nature, as an experimenter, in the matter of feeding. Experimental Nature shows us that the sun feeds, and that no experiments, conducted in a cold climate to show the vivifying effects of warmth on starving animals, are called for. Nature, in another experiment, shows that the sun, by an intense heat long maintained, can produce, under very slight animation, febrile heat in a warm-blooded animal, and kill, by increment of heat, with tetanic muscular rigidity. There are, in fact, no experiments more cruel, none more decisive, than those she performs without any invitation or help from us. She bakes and she freezes animals in her experimentation, and leaves her work behind her for our contemplation beyond any amount we can ever wish to see, still less produce, of such phenomena.

The study of the influence of external conditions on life and on living things affords, then, the widest scope for the student of physiology

that he can possibly require. The universe is
an experimental laboratory, in which he may
walk from table to table, from earth to air or sea,
and never tire, and never fail for subject of dis-
course. He may be satisfied, and he often only
cramps himself when,

" With all appliances of art to boot,"

he enters a laboratory of mere bricks and
mortar for pursuit of vital experiments.


The Zoological Gardens of Regent's Park beat
all the physiological laboratories of the world
put together. Every time we walk through
them they suggest new work of the highest, most
useful, most painless, kind. Do fishes sleep? A
friend of mine discussed this with me as we
studied the aquarium. We neither of us knew,
but here we could learn. If fishes do not sleep,
or if some do not, what is the reason of the fact?
Here we might learn. What are the products
of respiration in fishes, living, as they do, with
little oxygen? Here we might learn. Why do
fish, when Nature freezes them in stagnant
water, sometimes thaw and come to life again,
even after long and hard freezing, or sometimes

thaw, and show no sign of vital restoration ?
Here we might learn.   Indeed, what might we
not learn from careful study of natural experiment, and always with enlarging views, in this
living laboratory ?

There are many smaller laboratory studies
that could be carried out without the slightest
pain, and which promise to the investigator
the most fruitful return for industry.

The hydration of tissues is one of these later
studies to which in illustration I may draw
attention.   Dead tissues, muscle, brain, and
glandular tissue, subjected to hydraulic pressure
in order to see if, under such pressure, the water
contained in the tissue can be directly expressed,
gives a study.   This experiment teaches that
some tissues, at least, from the water with which
they are charged, act under pressure simply like
water itself ; they are incompressible so long as
their gelatinous or colloidal structure remains un-
decomposed and hydrated, in which state water
cannot be directly expressed from them.   But if
to the tissue, charged with water, saline substance
be added, expression under a comparatively light
pressure will take place.   The saline matter

seizes the water, fixes it, and, if channels be given for its extrusion, lets it·pass. In this manner we make, as it were, an artificial gland ; and by the research on dead matter seem to have arrived at two vital facts—that saline material is necessary for excretion by glands, and that saline material present in the body itself, during life, plays this physiological purpose. The salt made in the body—a cyanogen salt, urea, salt of the urine, the substance by which the nitrogenous part of the body is carried away—fixes and carries off under the blood pressure the water of the kidney, as if there could be no such act as excretion of urine in the absence of urea. The salt of the urine seems required to fix the water of the kidney, before the water, under the pressure of the blood, can be carried away by the uriniferous tubules into the pelvis of·the kidney and so through the ureters into the bladder. The same appears to me to be the case with all the other secretions ; each one must have its saline substance, by which its water must be extricated. The colloidal part of the organ must be hydrated, then, by virtue of saline matter, de-hydrated, as

explained above. Salines, in this manner, sustain
a vital current, and prevent what would otherwise be universal hydration of colloidal tissues—
universal dropsy. The value of salines, administered to produce free elimination, by the skin,
kidneys, and bowels ; as diaphoretics, as diuretics,
as purgatives, has, in this manner, a ready explanation.

I notice this research because of its immediate
practical bearing; but it is one of many and many
inquiries that wait for experimental elucidation
through inanimate matter. One of my masters,
the illustrious physicist Dr. Thomas Graham,
whose original labours on osmosis and hydration
were amongst the most remarkable of a physi-
covital kind that the present century has produced, left a mine of wealth from researches bearing on life and living action, all performed on
dead organic material. The dialysing power of
every membrane of the animal body has yet to be
determined ; the separation of tissues by dialysis
has yet to be determined ; the communications
of movements of cells have yet to be discovered,
and the break up, and the reconstruction, of cells
has yet to be resolved. Were I to extend sugges-

tion as I see the path open in the direction here shadowed forth, I should fill a large volume, for the synthesis of life itself lies here. I must at this moment rest, as an index finger, nothing more. I must let organic chemistry pass with the mere name, and leave its splendid and little acknowledged field of researches with a word.

I have not written this chapter with the idea of condemning all vital experimentation. There is a class of painless experimentation on living animals which, judiciously and properly conducted, may render useful and expedient aid in solving difficulties suggested by the experimentation of Nature. Against perfectly painless experiment, carried out for purely exceptional and great objects by men who themselves regret the necessity or expediency, and who only act under a strict sense of duty, no reasonable or just mind can raise an objection. For my part, I would object as earnestly as any one against the performance of minor and curious experiments, in which there is any infliction of suffering; I would object to all experiments implying suffering, which are nothing more than a repetition of the painful experimentation of Nature or of man. I am content to let Nature do all the torturing and man all the relieving; but I object also to the infliction of the torture of sentiment, put forth too freely by the friends of animals, and addressed to men who, under a solemn and sincere sense of duty, feel it expedient to experiment upon living creatures. What I am anxious to bring about is a reconciliation between extremists of both schools. I see evidences of reckless cruelty in both, and would it were removed! As for or against experimentation, the aim of all earnest men and women should be, not to breathe wrathful cruelty against each other, but to come to a common understanding.

The reform that is called for is the enlargement, or the widening, of the boundaries of physiology; the turning of crude physiology into accomplished biology; the destroyal of the absurd fashion that has grown up amongst physiologists of looking with a kind of scorn on all physiology that is not vitally experimental, and with contempt on all physiologists who avoid experimenting on living animals. Plainly, what is wanted in practical physiology is not more knowledge, but wisdom, with the strongest leaning towards all that is most humane. I may not go with some who contend in argument for the abolition of all vital experiment, but I quite agree that the grandest physiology and physiological discovery could exist outside every shade of painful experiment, and I am now as opposed as any one to methods of research that would take a living animal to pieces in order to discover its mechanism as if it were a watch, and means which, in the case of a watch itself, would be rude and ridiculous attempts for purposes of discovery.

# CHAPTER IV.

## CAUSES OF DISEASE, AND TREATMENTS.

QUESTION 4.—"*In the study of causes of disease and successful treatment, what sound methods of research can be laid down without recourse to painful experiments?*"

IN order to answer this question in all its fulness it would be necessary to go back to the oldest known records of medicine, and to trace the course of advancement without experiment from the most remote, until the present, time. It is well within the memory of practitioners still alive that the original study of disease and its successful treatment were entirely carried out apart from all idea of experiment in any systematic form. It is indeed only within recent years that we have departed from the old lines of observation, and entered upon the new, with experiment as an aid to research.

It is rather crucial to ask me, one of the early

workers of the new and experimental school, whether sound lines of practice can be laid down without experimentation ; but I do not hesitate to affirm that there are such lines, and I have no difficulty in explaining certain of them which must be sound, and which may perchance be sounder, in the long-run, than any rules laid down on a theoretical and experimental foundation. I am glad, in fact, of the opportunity, for I cannot be blind to the truth that the study of disease and its treatment is becoming too experimental, to the exclusion of the old and recognised methods in which the seniors, amongst even ourselves, were educated.   There are many lines of observation open for a renewal of research on the old plans which the modern student and practitioner cannot do better than follow.

There is a wide field open for the discovery of the every-day and natural causes of disease. Let us begin with the most elementary of studies, a common cold.   Nine patients out of ten who consult the physician respecting acute affections of the lungs, or about affections in which the lungs are not immediately involved, begin by relating that the symptoms first appeared as a cold. The physician, on his part, as frequently opens by asking whether a cold was not the first event, if the sick person does not supply the information on his own account. But what is a cold? What is the direct influence that leads to a cold? What is the best means for removing the symptoms when they appear? There is no answer to these questions that is at all reliable. I have been warned, ever since I can remember, never to sit in a draught for fear of taking cold, and the warning I know full well to be wise and prudent. But what has sitting in a draught to do with the seizure of a cold? There is no answer yet to that simple question. Some sound rule is wanted here, based on some simple discovery which no painful experiment on a lower animal is likely to reveal. That which is wanted is a precise knowledge of atmospherical variation. What occurs when air strikes the body in a current and so affects one side of the body. A sound observation on this subject would be one of the most practical of discoveries ever made, both for cause of disease and for treatment. We are not in possession of what Dr. Cophagus called the "rudimans" of physic

while such an innocent question remains un-
answered.

Nature in this case presents an experiment
which she has performed millions upon millions
of times, an experiment which she has probably
performed on every son of man that has been
born. Yet so badly have we read her work we do
not follow her in so simple a diversion. Nothing
is demanded except industry of observation, bearing not on the subject who takes the cold but on
the surroundings which lead to the development
of the phenomena. The student of medicine
who will devote the best years of his life to the
investigation of this one subject; who will call
chemistry to his aid, meteorology and physical
science in its special application to external
influences affecting life, has a field of painless
research before him that will bring him the best
reward, and may crown him with brilliant
discovery.

I never enter the wards of the hospital to
which I am physician without knowing that I
shall find there one or more cases of acute
rheumatism. Sometimes I have so many cases
in different stages it would·be possible to demon-

strate, briefly, to a class, every phase of rheumatic
disease in one afternoon. I rarely enter a family,
as a medical man, without hearing, if I make the
inquiry, that some member of the family, past or
present, has suffered from rheumatism. I cannot
take up an ancient treatise, to say nothing of
modern treatises, on the practice of physic
without finding descriptions of rheumatic
disease. I read of them in works of foreign
countries, or I speak with medical travellers
about their observations. I find from this information the story of the universality of this
scourge on the earth. I find it told that some
diseases interchange positions, that typhus fever
ceases at a thermal line where yellow fever com-
mences; that this place, or that, is free altogether, or is specially affected with some particular malady, but no place seems to exclude
rheumatism. It is present in arctic, it is present
in torrid, it flourishes in temperate, regions. We
dread it here in cold weather and wish we were
in equatorial climes so as to escape it; but the
native of Bengal, whom we may upbraid because
he eats or smokes opium, will explain to us
that his habit has been engendered by his desire

to overcome the rheumatical malarial suffering to which his climate subjects him.

Rheumatism, then, is a disease truly universal. It spares neither rich nor poor; neither northerner nor southerner. It afflicts young and old; it is everywhere and apparently under the most diverse conditions in which the human family exists.

The study does not rest here. Rheumatism is not only rheumatism marked by fever, by red, swollen, and painful joints, and other acute symptoms. Since Sir David Dundas first observed the connection of heart disease with rheumatism we have learned, clinically, how the serous and fibrous structures of the body may be injured by an acute rheumatic attack, and what crippling mischiefs may follow. Heart disease and rheumatic disease are largely synonymous terms with us now. An insurance company never reads of a candidate for insurance having been the victim of acute rheumatic fever without inquiring whether he has or has not disease of the heart. Physicians are not content to rest at that point. They look at other diseases in which the membranes are implicated, such diseases as chorea, for example, and neuralgia, originating in rheumatic disorganisation ; and, not to linger longer, rheumatism may, indeed, be accepted as a base of more chronic disease than any other complaint, and, in the acute and chronic form, is perhaps the most universal affection afflicting humanity, and afflicting also many species of animals inferior to man.

All this burden of disease has one common origin. There is no such thing as two primary origins possible. What is the origin ? What cause is it that diffuses itself so widely ? It must be something very simple, something very easily controlled. Is it diet ? Is it dampness of air ? Is it stamp of heredity ? We have by experiment—I have myself—produced a kind of synthesis of the phenomena in susceptible animals, but the attempt went a very little way towards explaining cause and course. It showed that a chemical product of fermentation, being put into or generated within the living body, was capable, as a secondary product, of producing the symptoms acute and chronic ; but it failed to indicate how that product which the rheumatic patient makes and eliminates during the disease is manufactured in the organism. It failed to tell what conditions lead to the perverted animal chemistry from which the exciting product springs. All bearing on that subject is left for discovery to that student or to those students of Nature who will most carefully inquire into her experimentation in the production of rheumatism, and who will circumvent her in her unmeasured cruelty.

Sir Edwin Chadwick was accustomed to say that good drainage would cure toothache. Those who did not understand him laughed at the saying, as at the crank of a philosopher. Yet he may have been quite right; and one day some man of science may prove that a similar simple remedy may save the whole world from all the misery and fatality arising from rheumatism. There is the natural experiment, any way, going on before our eyes : let Nature inflict the pain, let Science bring the relief; let Science find out the conditions under which Nature works to produce the symptoms, and the victory will be with Science; for be it understood that Nature cares not how things go ; she is neither merciful nor vindictive ; she merely pursues her definite course, irrespective of results, good or bad, to life and living action.    There is grand scope for inquiry, without any painful experiment on lower animals, in the disease rheumatism.    The student who would enter into it need not trouble himself about effects following the acute or primary manifestation.    His business is with the primary manifestation itself, and with those circumstances which give it birth.    Historically, he has such work as Dr. Creighton has shown him in the design of the magnificent volume on epidemics.    Meteorologically, he has all the field of observation bearing on the effects of atmospherical variation, of which Dr. Weir Mitchell has afforded such a model in the essay on neuralgic pain as determined by changes of atmospherical pressure.    Chemically, he has the investigation, which Dr. Marcet has ably opened, on the products of respiration in the course of disease, together, with other chemical inquiries into cutaneous, renal, and intestinal excretion.    Statistically, he has the collection of a perfect harvest of facts relating to the occurrence of rheumatic affections in different

climates, in the different sexes, in different oc-
cupations, and in the different periods of life.
Geographically, he has the study of the affection in various terrestrial positions, after the
lines so laboriously laid down by Mr. Alfred
Haviland. Clinically, he has the most extensive studies under his hand, in the range
of symptoms, human and comparative. Patho-
logically, he has endless museum specimens
for examination.

I have named two diseases; one a meteoro-
logical disease, the other possibly a zymotic,
calling not for experiment beyond clinical observation. But I could multiply the illustrations to a wide extent. They could be made
to apply to most diseases with which the human
family is oppressed.

No reference has been made, so far, in this
chapter to what is experimental in the in-
offensive meaning of the term. The omission
does not carry with it the intention of exclud-
ing experiment as a means of research, for, in
truth, there is any amount of experiment before
the inquirer who wishes to avail himself of it

as a means of progress in discovery. We
ought to disabuse our minds altogether of the
misuse of the word. They who are opposed
to vital experiments are so intense in their
expressions they forget to be cautious, and in
many instances oppose experiment on the
sound of the word alone. They, on the other
hand, who are warmly in favour of experimental pursuits, are led by a kind of impulse
of opposition to attach too much importance
to experiments that are vital, and too little
to those which are simpler in their character.
They who are impartial will draw the line at
painful experiment, and will give to experiment
that is painless all the immense importance it
deserves, and leave the inquirer free to carry
out a world of work to which, really, there
seems no end. The whole question of motion
excited by zymosis or ferment is in this
manner left, and a wide and profound subject
it is. Whatever is worthy of good report in
the hypothesis, founded on analogies and co-
incidences, called the germ hypothesis, or,
incorrectly, the germ theory, is open to investigation, by experiment, that the most sensi-

tive need not object to. The actual nature of
the force of life itself may lie open here. Life
is, practically, zymosis, and inanimate matter is
made to move as if by its own act and endowment. The motion we call living motion—
with or without sensation—may be fermentation. There is as clear manifestation of life, in
mere motion, in the fermentation of beer, as in
the wrestling of a giant or the flight of an eagle.
All may be from zymosis. We hear of zymotic
diseases, as if, in communicable diseases attended with fever, there was set up some new
process in the living body which may kill,
and which has nothing to do with the normal
life ; whereas the febrile excitement is as clear
a continuance of the natural process of life as
a huge fire is the continuance of a moderate
fire burning high, for a short time, on the
addition of more inflammable fuel or more
oxygen.

On the whole matter of animal zymosis the
student has scope for experimental work, and
for applying his work to the study of disease ·
and its successful treatment. He may devote
his energies, with every certainty of obtaining

reward, to the cause of zymosis as a pure
physical phenomenon. Why does the presence
of some minute form of organic matter, itself
vital, move other matter into the state of life,
and from one minute point generate a world
of life just as surely, if the circumstances admit,
as the spark from a flint and steel may generate
a fire that may burn down a city? The student
may follow up his research. He may start a
zymosis akin to the animal zymosis, and,
without calling into play any sensitive part,
may follow up all the variations of zymotic
action, all the perturbations of it, from inter-
ference, as well in a senseless as in a sensitive
organism, and may apply every test he wants
for elucidation of animal perturbations equally
well out of the animal body as in it. By
this plan he is likely to lead to better results than can be obtained by making similar
inquiries on sensitive structures, in which the
added sensitiveness is liable at every turn to
falsify the physical manifestations incident to a
condition so purely physical as motion is incident
to fermentation. It is quite true that in the
conduct of the research which the student would

pursue in these studies he would find at first
a profound difficulty of apprehension and a separation of what seem to be combined actions in
the living body. He would have to separate,
in his mind, the force of living motion from
that of living sensation. The forces are distinct, and the fact of the distinction must be
recognised. Physical vital force, motion of life,
is fed by chemical combinations and decom-
positions ; sense, sensibility, nervous force, is
fed, through the senses, by vibrations from
the outer universe. The first is from within
outwards ; the second is from without inwards.
The first is structural, of the earth, earthy,
variable ; the second is structureless in force
or forces, and always acts healthily when the
matter on which it vibrates is normal, but
diverts strangely when the matter which it sets
in vibration is not normal or is obscured. In
the dark I am blind, and blindness is a disease ;
yet I may have no disease. Under cataract I am
blind in the presence of ordinary light, because
I have physical disease of a structure, or lens,
that fails to transmit and refract ordinary light.
The sensibility of vision and the disease are two

distinct things, and I need not consider the sensibility in having to clarify one part of its physical mechanism. I might learn how to clarify a dead opaque lens, with the best results in respect to what would be learned as to treatment of cataract without operation. The same method I believe to be universal in application, even in application to pain, which, as aberrant vibration is all but certain to divert falsely, if it be promoted in the course of experiment, in sentient matter. The more, indeed, the mind dwells on the observation of vital phenomena under pain, the more it must doubt. A patient I knew, suffering the acute agony of passing a gall stone, not knowing what else to do, stood on his head. What if that had been chronicled as a symptom of gall stone! Yet there are records of results of experiments under pain which are quite as inconclusive.

To sum up, while we need not exempt inferior animals from rational experimentation for some important purposes of research, pain, as a disturbing influence, is of so serious a character, that, quite apart from sentiment on the matter, it were, I think, best to exclude it altogether. It cannot guide; it must deceive. I doubt whether there is one experimental statement announced through phenomena of pain that is reliable. No two human beings give the same definition of the pains to which they are subjected. No two women give the same account of the pains of childbirth. How, then, can signs of pain in experiment be booked with any accuracy by an observer who does not feel them, or yield any practical result?

Reference has been made above to the study of disease, and similar reference might be offered on treatment. There is the most ample scope for the study of treatment without painful experiment that any one can desire. Treatment at this time is perfect Babel. Every man prescribes after his own heart, and holds to remedies so entirely his own, that two men scarcely ever write the same prescription for the same disease or for the same symptom. I have watched the art of prescribing for fifty years, and I am quite sure that divergence of treatment is at this moment far greater than it ever was in the course of that long period. The multiplication of remedies, begotten of experiment, is the chief reason of so much disagreement. No remedy now holds its place except for a short season. Each comes and goes like a meteor. Some sick persons make pilgrimages to cures ; more frequently cures make pilgrimages to the sick, and the results at this unsteady stage are much of a muchness. Altogether, there is far too loose a system of medication. The modern student has before him a new duty. The experiment of experiment that lies before him therapeutically, is to learn what diseases will recover by mere attention to external conditions without any medicines, and what will not. He may in this research, bearing on some acute and, at present, fatal diseases, be able to put the sick in such conditions that they cannot die from their disease ; but this supreme art will never come from the current use of drugs in special form. To prescribe for pneumonia some new or old remedy, and leave to any one the care of the fire and the window of the sick room, may be practical, but it is not progressive, nor well calculated to cure.

To administer the same medicine without reference to season, or dilution, or connection
with food, may be practical ; but it is not
progressive, and experimental researches for
remedies have not much chance under such
irregular methods of administration.

It cannot be long that this chaos shall
remain. The world will force the professors
of medicine into a new path if the professors
do not find it for themselves. When the right
path is found, diseases will be divided definitely
into two natural classes; those which require
no special treatment of a medical kind, but
which get well of themselves under continuance
of natural law ; and, those which, under continuance of natural law, will progress as diseases
unless man interferes, and by his skill averts
the fatal catastrophe.

I have said that the first grand experiment, the
experiment of experiments, consists in making
the classification of diseases into the two
classes, the naturally curable and the artificially
curable. Experiment turns on the latter
alone ; in what is incurable, except by art,
experiment is the only method at command

for advancement. We cannot, therefore, give up experiment in treatment, unless by ceasing, altogether, the attempt at cure. Much can be done by experience alone; some, as far as I am able to discern, can only be effected by experimental pursuit, which, however, need not be painful.

At the risk of being too much in evidence through my own researches, I will explain what I mean on this point from two of them. I shall in this way not only speak from personal knowledge on methods of advancing treatment, but shall take on myself all the criticism which the methods adopted may call forth. I once introduced, for reasons of a chemical and of a curious kind, a medicinal substance known by the name of sodium ethylate, or caustic alcohol. Common alcohol is made up of three elements in different equivalents. Two equivalents of carbon are combined with six of hydrogen and one of oxygen. If now, into a portion of this compound fluid we put a portion of the element sodium, in its pure metallic state, we observe at once a commotion and an escape of a gas which, being collected and examined, proves to be hydrogen gas. Hydrogen escapes, the sodium dissolves and disappears. In place of the white mobile fluid, alcohol, into which we dropped the sodium, we have produced a dark, rather thick fluid, which will crystallise with greater weight and with other qualities differing from alcohol. In the change that has occurred we have had a process of substitution. A part of the hydrogen, the part which escaped, has been replaced by the metal sodium, so that now the compound contains two parts of carbon as before, with one part of oxygen as before ; and, instead of six parts of hydrogen, five parts of hydrogen and one of sodium. But sodium is a metal greedy for oxygen, and, combining with oxygen, forms soda, or oxide of sodium. The whole forms sodium ethylate, to which, if water be added, another change or reconstitution takes place. Water is composed of hydrogen and oxygen. The oxygen of the water goes over to the sodium, forming soda, while its hydrogen, liberated from the oxygen, takes the place held originally by the hydrogen in the alcohol, and alcohol is re-formed. This, without minuter details which might confuse readers who are not acquainted with the laws of chemical combination, is the chemical part of this process; and it led to the medical. Soda is a potent caustic and destroyer of living animal tissues. Alcohol is greedy for water, and, added to animal tissues or fluids that are coagulable, such as albumen, produces instant coagulation and 'solidification. But soft animal tissues are rich in water. What, therefore, I reasoned, would occur if they were treated with sodium ethylate? I tried the experiment on coagulable dead fluids, and found, as anticipated, that sodium ethylate would decompose them and solidify them. The transition of experiment from a dead to a living structure was easy. A raised vascular tumour or nævus on the skin of a living person was treated by sodium ethylate. At once the water in the nævus was decomposed. The sodium combined with the oxygen of the water, making caustic soda, and the hydrogen, re-combining as originally, reproduced alcohol in the nascent state. The soda thus formed acted as a caustic or destroyer of the animal tissues, while the alcohol seized the water of the tissues and coagulated them, converting them into a hard dead substance, vulgarly called a scale or scab. A portion of the nævus was thus superficially destroyed ; by repeating the process the whole mass was destroyed, and, what is called a cure was scientifically effected by following out a purely physiological process.

To such a method as is here described no one could raise objection, for no experiment on an inferior animal was inflicted. Yet it is not less certain that the result obtained was a product of experimental research. I should never have approached the result by any other line of inquiry ; it came from thinking of the principle of the chemical constitution of reme- dies in respect to their applicability in practice for remedial objects ; and, if I did not need to experiment on an animal in this instance, it was, firstly, because no danger to life was involved in the treatment, and secondly, because, as from experimental knowledge I was certain of what would happen, it would have been a work of supererogation to have performed an experi- ment on a lower animal.

It is not always that we are brought so readily,

by an experimental process, to the study of problems relating to cure. Some remedies which,
on the physiological argument promise the best
results, are too dangerous to be tried at once
through the disease, and with them, therefore,
preliminary experiment on a lower animal has
been considered essential. I will relate, as an
illustration, what may be called a necessary
vital experiment.

We treat now with much success some of the
most excruciating diseases incident to humanity,
such as colic, tetanus, and angina pectoris, by
a remedy, the virtues of which, up to a certain
degree, are beyond dispute. This remedy is
known as nitrite of amyl. The remedy came,
originally, into my hands for investigation in
the year 1862, with a very brief history, to the
effect that the late Professor Guthrie, in dis-
tilling it, had found that the vapour of it
produced, when inhaled, flushing of the face,
excitement, quickened pulse, and quickened
breathing. When the nitrite came under my
observation I commenced by breathing the
vapour of it myself, and produced a train of
symptoms of the most potent kind. My pulse

rose to one hundred and forty whilst breathing the vapour ; my skin assumed the redness of the skin in scarlet fever, and my breathing was greatly accelerated. Several other persons, at my request, inhaled the vapour, and observation on the human subject was carried to the fullest degree that was compatible with safety. Out of the phenomena observed nothing came that was satisfactory. All the phenomena were chaotic, and, for anything I knew to the contrary, they would, if carried further, terminate in sudden death. I determined, consequently, to carry on the inquiry by subjecting inferior animals to the vapour up to the extent of producing the most decisive and, it might be, fatal conditions. The experiments were made both on warm-blooded animals and on batrachians, and from them came out two of the most astounding facts I ever had witnessed. First, it was discovered that, from the action of the nitrite vapour on the pulmonary vessels of warm-blooded animals, variations in the circulation of the blood through those organs could be brought about, ranging from bloodlessness, or syncope of the lung, to the extremest congestion, showing that this chemical body had the property of making the heart completely inject the blood vessels of the minute pulmonary circulation.

This was an important fact, but it was eclipsed by another. Some frogs that had been subjected to the action of the nitrite, as we human animals had been, but in a more fatal degree, apparently died without convulsion or any visible suffering. In this state they remained entirely flaccid ; that is to say, they never showed any indication of that rigidity of the muscles after death to which we give the name of *rigor mortis*. They were laid in moist moss, and were observed from time to time for many days in order to see if *rigor mortis* (death rigidity) would come on, or whether they would pass into decomposition without showing any sign of that process. It is the fact that some of them showed signs of commencing decomposition in the web of the foot, a sufficient indication, as I thought, of absolute death, when suddenly, at the moment when the animals were about to be consigned to the dust-heap, motion was observed in the limbs of one of them. The phenomenon was astounding, but more remained. These apparently dead creatures came to life, and regained all their vital activities without any obvious suffering.

From these observations I was led to a definite conclusion about the action of amyl nitrite ; from them what had before been mere chaos of thought was clear thought. This agent, I said, acts on the nerve fibres governing the circulation like a physical injury, such as the division of a nerve which paralyses nervous function. I wrote at the time in a report to the British Association, in 1863, that here was a chemical substance which produced on nervous function the effect of a physical injury. One further inquiry was forced on me. Would this chemical body, which possesses the power of suspending the contraction of muscle incident to death, relax a muscle that has been made to contract spasmodically ? The trial was conducted, and it answered affirmatively. Nitrite of amyl, by its powerful relaxing effect, would relax even a tetanised muscle.

The therapeutical worth of the nitrite was now laid open, and I reported on it " that it was the most potent antispasmodic that had ever been discovered." It was a remedy we wanted for the acute and most agonising of diseases,

human and comparative; it could be used with
hopeful effect even in tetanus itself. The pre-
diction was fulfilled. As amyl nitrite became
applied for tetanus, for angina pectoris, and for
colic, not to name minor diseases, its effects
came out so decisively that a new step in the
treatment of those affections was taken, and,
with various and important additions of knowledge from other observers, it continues to be
employed to the present time.

In this instance vital experiments on the
inferior animal were unquestionably of service,
and the probabilities are that in their absence
the results discovered would have been lost.
The first symptoms produced by the nitrite
on the human subject were so unlike those of
an antispasmodic and sedative, and were so
extreme, they suggested the dread of death
from spasm of the glottis, and from acute spas-
modic action of the heart. It chanced truly
that a little later on, a medical friend, calling
on me, and finding me at work with the nitrite,
inhaled, during my temporary absence, and, in
sheer incredulity of its effects, so large a dose
of the vapour, that when I returned he was

prostrate on the floor, his whole muscular system so relaxed he could not move a limb, in which state he remained painlessly helpless for several hours. But this, which might have opened my eyes, was such an alarming accident, it made me too anxious to allow of any calm deductions from it, and might never occur again in a lifetime.

Experiment, therefore, came in usefully in this discovery—if I may use the word—of the action of amyl nitrite. It was not perhaps indispensable, but it was very expedient. It led also to other valuable inferences and applications. The flush and excitement of the circulation, and the muscular collapse following, which are produced by alcohol, came to my mind, and caused me to read, I think correctly, the modus of alcoholic intoxication. The explanation extended to the action of atropine and other substances, and finally brought me to practise the method of inquiry into the influence of medicinal substances by commencing with an elementary base, and following the modifications of its action through the varied compounds formed upon it. That was a new method of study in therapeutics, and upon it all my therapeutical work for many years rested. Previously observers had tested remedies in their gross forms ; tested them empirically by experimenting with them directly on mankind through disease. Some accident led to the first trial, as when water, charged with something derived from cinchona trees, was seen to cure ague. Now we test remedies by their construction. All remedies are made up of certain well-known chemical elements, each of which has its own influence on the animal body. How will any element be modified by combination with one or more other elements? We may, perhaps, follow up every part of this inquiry by direct experiment on mankind through disease, but it seems expedient at first to ask many preliminary and vital questions through animals inferior to man. It may be so in the future, yet the experiments need not be painful.

I may close the answer to the comprehensive fourth question by saying that the study of disease and of its successful treatment may run on the most advanced as well as on the most practical lines of research without painful experiment, and in the main, without any experiment at all on inferior creatures. It would be easy to give many proofs of this last statement, and I will offer one again from personal knowledge. The medicinal fluid called hydrogen peroxide was originally studied and brought into use by myself in 1858-59. It is now one of the most largely and successfully used of medicaments. It was investigated altogether, from its first introduction, apart from any painful experiment on any animal body.

# CHAPTER V.

## THE NATURAL METHOD OF PREVENTION.

QUESTION 5.—"*What do you recommend as the best methods of obtaining the great aim of medicine—namely, sanitation and the prevention of disease?*"

YOU might expect from me, in answering this question, a long dissertation. It need not be so. My mind has for many years past been made up on this matter, and I could put all the answer into the one word, *cleanliness.* If this, or any other country, could rise one morning with the sun and find itself perfectly clean, the great aims of medicine—namely, sanitation and the prevention of disease—would be realised, so long as the cleanliness was maintained. Whenever, therefore, I assist in the work of sanitation, I feel that I am almost doing a work of super-erogation in the carrying out of duties not more profound than any good housewife might perform—the best type of a health-giving service.

By cleanliness, however, I do not mean housewife or washerwoman cleanliness alone. I would bring under it mental cleanliness also. It is good to apply cleanliness to the health of the body, but it is an imperfect effort to carry it out unless we make it also apply to the health of the mind. There is one physical and mental disease afflicting mankind in every clime. It is the disease of human rottenness and lust. It is the curse of foul humanity. It passes from generation to generation, involving in its loathsomeness the innocent and the guilty. If it were removed half the constitutional disease of the world would cease. We call it politely and forcibly "specific disease," but all the doctors in the world will never cleanse the world of this disease, by mere physical attempts to cure or prevent it. It is a disease of the mind as well as of the body ; and until the mind is cleansed of it the body will remain degenerated by it. Specific disease is a startling type of all diseases that have to be eradicated, and of the means for their eradication. It is a contagious disease; it spreads by inoculation of its kind, and by no other general way; no one need have it unless they allow themselves to be inoculated by it. It is cultivated by human beings as living poisonous plants might be, and it is under as complete control as a plant might be, and is as capable of entire extirpation.

Every kind of remedy has been proposed for this disease; every kind of means has been carried out for its prevention except one, *purity*. Wise creatures carrying aloft a filthy rag of science bestained with the disease have proposed, and have practised, the attempt to render the human family exempt from it by a process of artificial infection, as if it were not foolish enough to get it by inoculation in what is called the natural method. Still the specific disease goes its way, no one checking it in the least by their artificial attempts at prevention or cure; and it will go on mocking science mocking art, until physical and mental cleanliness unite; until the mind as well as the body is clean.

I give up all ideas of physical experimentation for the entire purposes of sanitation, and prevention of disease. They are not, and never can be, basic. They are not, and never can be, wanted. If they succeeded they would be false in action, for they would bind liberty and progress with bands stronger than iron. They do partially succeed by violence of method. We lock up certain of our populations in experimental fortresses called prisons ; we force them into sanitary conditions, and the experiment suc-ceeds. Prisoners immured in prisons are prevented, *nolens volens*, from taking the diseases sanitation aims to prevent. The object lesson is perfect ; it shows what cleanliness can do, and it shows that, when it governs, sanitation and prevention succeed without other artificial methods of any kind. Who would think of inoculating prisoners to prevent them having " specific disease," or any contagious disease, so long as they are separated from contagion ? Who would not feel that to do such an act would be to introduce disease into the centres of enforced purity ? One would hardly think of vaccinating a prison community for the prevention of smallpox even though it were susceptible to smallpox.

I recommend, as the best method of obtaining the great aims of medicine—namely, sanitation, and the prevention of disease—first, to make medicine the grand master and teacher of universal cleanliness, and to make every one of the community a disciple and follower of the same law. The minister of medical art should be prepared to devote his life to this simple duty. He needs no higher calling, no nobler vocation, and a world that knew its own interests should sustain him in his task. This should be the *esprit de corps* of physic.

At present the rage is for experimentation, although it seems least wanted, for which rage the selfish and ignorant world is most to be blamed. The world now, as in the days of Naaman the leper, wants to be healed and protected by elaborate processes when the simplest and surest remedy is in its own hands. In 1881 some one proposed a scheme for having the domestic cattle of England inoculated with a fell disease in order to give the cattle immunity from that disease. At the moment one of the leading practitioners of the veterinary profession, who had also one of the largest practices, told me his belief that there was not a case of the kind in the length and breadth of the land, and that during his long and wide experience it was the rarest disease known to him. So a most formidable disease was to be inoculated to prevent what did not exist. Towards the close of last century a census was taken of the town of Brighton in order to arrange for the universal inoculation of the smallpox to prevent the incursion of small-pox. Could human folly go farther? Think of the art of making every house of a fashion-able watering-place and centre of health a hotbed of a contagious disease. Some of us stand in astonishment at this time when we hear of so filthy a proposition; yet there remain thousands who, in many ways, are re-peating the method, and are prepared to sustain it as a principle of action, rather than " wash, and be clean."

. I support, heartily, for sanitation and pre-vention of disease the one natural system, in which every man and woman may take part—the system of external and internal purifica-tion. There is an ancient Talmudic proverb

9

in the *Mishna* which says " Outward cleanliness is inward purity," a proverb which in our day has been rendered by John Wesley into " Cleanliness is next to godliness."   In a world of disease, caused by sheer uncleanliness, cleanliness is the lesson to be enforced on the mind by precept and by example from the earliest days of life.  Let us drain our country on such a plan of uniformity, that every particle of pollution shall pass from our houses as it is produced there ;  let us cleanse our outward garments, our bodies, our food, our drink, and keep them cleansed ; let us cleanse our minds as well as our garments, and keep them clean ; let us isolate the contagious sick  as  they become contagious.   Then all elaborate experiments for the prevention of disease by artificial methods of production of disease will appear, as they are, mysterious additions to evils which ought not to exist, and which of themselves might re-introduce death into a deathless paradise.

# CHAPTER VI.

## ERRONEOUS RESEARCH.

QUESTION 6.—*" Do you agree with us in the view that both intellectual and moral evil result from erroneous methods of research?* '

EVERYONE, whatever the opinion is about experimentation, must, I think, be in favour of the view that intellectual and moral evils result from erroneous methods of research. The question is :—What is erroneous method? A repetition, in other words, of the inquiry of Pilate, "What is truth?" If all that is done in the way of vital experimentation be erroneous, then the dissemination of so-called knowledge on that subject is a cause both of intellectual and moral evil. It is a curious fact, which I do not remember to have seen recorded, that every method of research that is most enduring, most intellectual, and most free from moral evil, is

plainest, and is furthest away from any and
everything that shocks the conscience or raises
a doubt, as to necessity, in sensitive minds. The
strength of mathematical science, the charm of
it, lies in its freedom from causing sentiment
of pain. If mathematics had to be cultivated
by and through experiments on living animals ;
if it led to the shedding of even drops of blood,
it would never have succeeded in unfolding the
magnificent mysteries of the universe it has
laid bare to the common intellect by the labours
of the godlike men who have been its masters.
If its professors had declared for it the necessity
of vital experiment, and of practising experiments on lower animals, they would have created
a disgust that would have interrupted their
career in ages much less cultivated than our
own. The same applies to the work of the
sciences of chemistry, of botany, of mechanics,
and of physics generally, as well as to that
refined science which, completed, rises into per-
ception as art. In my opinion every man who
studies natural things by experiments on living
subjects, of any species, feels the truth of what
I am saying. I know, in my own case, that my

mind during such experiments has always been in a different state according to the line of experiment that has been under observation. When the experiment has been conducted on dead or inanimate matter, the return obtained from the labour demanded has always been, not only satisfactory, whether affirmative or negative in its character, but pleasant to the mind. On the contrary, when the experiment has been conducted on living or animate matter, the labour, whether affirmative or negative in its results, has never, at any point of it, been pleasant. The results yielded from living matter may, and often have, excited curiosity; they may have seemed to be highly important ; they may have been important, and they may have opened the way to new inquiry, or even to new practice ; but they have never been free of anxiety, nor of a sense that, whatever came from them, there was something that was not right. I do not believe I am more sentimental than any of my colleagues ; yet I never proceeded to any experiment on a living animal, though, to the best of my ability, doing everything possible to save all pain, without feeling what I think is the proper expression, compunction. I am not alone in this particular. Every person whom I have met pursuing the same line of research has acknowledged, when questioned, the same sensation.

The late Dr. W. B. Carpenter expressed himself to me strongly in that direction; so did Drs. Baly, Sharpey, Willis, and Snow. Fergusson, who was not quite consistent as between theory and practice, was still remarkably outspoken on the matter. The late Sir Thomas Watson, a man of extremely fine judgment, said that nothing but the sense of duty permitted him to witness the smallest vital experiment. Graham shrank from the ordeal. Richard Owen would never willingly witness experiment. " It made him ill." Brodie, in his later life, although he had experimented in his early career, told me that experiment on living animals should be exceptionally performed by men exceptionally qualified, men of true insight, who were able to resolve exceptional phenomena · or the causes of them. Sir James Alderson objected to any experiments on an animal in a lecture course. Dr. John Reid, of St. Andrews, in his late days was exquisitely touched in regard to the experiments he had performed. Sir Charles Bell was not disposed to painful experiments, and his cotemporary, Alexander Walker, loathed them.

I could name other men of distinction who have been of the same mind, but the above—many of whom I speak of from personal knowledge —are sufficient to prove that, what is vulgarly felt to be objectionable, may also be felt by the conscientious and sensitive souls of good men of medicine ; men who have no prejudices, but are as anxious as any to promote true knowledge in medical science, and any proceeding in it that would lead to moral and intellectual improvement. Objection is further borne out by the views which educated men and women in general feel towards the experimental brotherhood. I do not refer now to that anti-vivisection party which, in its vengeance, is led towards the infliction of the severest experimental punishment of a mental kind on those to whom it is opposed. I do not speak, either, as opposed to experiment ; because if I saw the possible discovery of an important truth lying behind experiment, I would vote for the experiment. I speak merely as a recorder of what I observe amongst men and women who are educated, just, and anxious to do what is good for the commonwealth individually as well as collectively ; and I can but record the belief that, even if good results seem to them to flow from physiological inquiries involving pain inflicted on inferior animals, they accept the benefit with reserve. It arouses in them no enthusiasm ; it comes to them not as the light of a new star, planet, or comet ; not as the discovery of a new land, river, lake, sea, food, fruit, or flower ; not as the application of a new force ; not as a new art ; not as an act of bravery, in which, though it be bloody, the courage of the combatant is put to the test ; not as any of these, but as if it were a cold-blooded cruelty which did not return any profitable interest, and which was to be tolerated merely because it promised, in the opinion of a limited number of scholars, to be prospectively useful. For these reasons experiment is confined to a limited number of inquirers, who speak, unless they be imprudent, only amongst themselves. Some, hard-working, honest, and resolute, brace them-

selves up to the belief—such is the imperious-
ness of their vocation—that, if a research has to be
carried out, pain should not stand in the way. To
these, research is not a matter of pain, but of
discovery at cost of pain, the interests at stake
not being measurable by pain, incident at the
worst to a few inferior animals—by no means
so sentient and sensitive as the human animal—
compared with pain which man endures from
accident and disease.

The argument would be unanswerable if three
parts were added to it. (1) If one experimentalist were as able as another ; had the same
insight, the same genius, the same appreciation
of the importance of any particular research.
(2) If the experiment were free from contradiction and error, so that opposition need
not be excited by it, nor vindication supported by it. (3) If there were in view a
time when experiment would have done all
that is required for the satisfaction of inquiry, so that the physiologist might close his
experimental history, in so far as researches
on living animals are concerned, with the
feeling that the epoch for such experiment

has passed, and that the ends justified the means.

A difficulty in answering the question lies in the way it is worded. I should have liked to have answered it negatively, on the ground that the method by experiment is not open to every teacher, is not open to error, is within measurable distance of its end, and is as little objectionable to the community at large as any other physical pursuit. It is impossible to admit that experiment on animals has the above qualifications. In the hands of the teacher it may be rankly abused; of scientific pursuits it is one most liable to error; it suggests no end to itself, but seems to grow by what it feeds on, becoming by repetition and contest more and more extended and multiplied; it is of all pursuits the most disliked by the educated community; it brings its best and most self-sacrificing professors into scorn; and for all such reasons, even if it be occasionally useful, is calculated to lead to what would be designated intellectual and moral evil.

These, in the general sense, are my convictions; at the same time let it be understood that I do not include in the criticisms experiments which are devoid of pain; nor experiments which, being devoid of pain, may cause even the death of an inferior animal for the service of man. Above all, I could not for a moment object to experiment well thought out by a truly competent man for the purpose of inquiry into some great theory that has been leisurely formed, and can be proved or disproved by no other means; as, for example, whether an important surgical operation can or cannot be performed for the saving of human suffering or human life.

It will probably be thought that the last-named method of experiment leaves the necessity for experiment in the same position as any other extreme proceeding that might be undertaken in an emergency, such as shipwreck, or famine, when what would be considered erroneous under ordinary circumstances is accepted as a part of the catastrophe, during which, in order to save human beings, lower animals are sacrificed.

# CHAPTER VII.

## INSTRUCTION BY EXPERIMENT.

QUESTION 7.—" *Do you approve of the instruction of students by means of experimentalism on living animals ?* "

THERE are some simple and painless experiments which may be demonstrated to any set of pupils, although living animals are the subjects of them. The demonstration of the circulation through the web of the frog ; the demonstration of the different natural temperatures of the bodies of animals, including man ; the influence of various anæsthetic vapours ; the collection of the breath of various animals for the purpose of analysis. These are all free from objection ; and it were hard to find fault with the process of taking a speck of blood for the intention of learning the various kinds of corpuscles which are present in the blood of animals of different species.

In a word, all experiments which are painless and harmless are, as I assume the most humane would admit, free from any charge of error in demonstration. But when we come to consider the application of experiment of a severe kind as a means of education of pupils who are making a study of biological and physiological problems, there is reason for hesitation. In my student days such experiment was never dreamed of. The professor of physiology would relate the facts derived from experiment, on which some important theories were founded; he would, for instance, explain what experiments were made by Harvey in order to describe the circulation of the blood; but he would not attempt to repeat those experiments in the lecture room. He would describe, in his remarks on the functions of the nervous system, the researches of Alexander Walker and Sir Charles Bell, in their original but contradictory observations on the functions of the anterior and posterior spinal nerves; but he would never think of repeating Bell's experiment of division of the nerves in the column, alleging forcibly Bell's own objection to its repetition. It was the same on every point. He would relate the theory; relate the *pros* and *cons*; relate, possibly, his own independent inquiries, or what he had seen experimentally performed by other independent investigators; but with that explanation he would be content.

Such was the state of affairs half a century ago. Gradually a new state has been evolved, and now some teachers demonstrate various views or facts by direct reference to experiment on living animals, on animals, like frogs, which have recently been decapitated, and thereby rendered insensible to pain. To these demonstrations women have been admitted, and if women are learning physiology, and if experiments are really necessary, the position is unassailable. The difficulty lies in the question, not whether women should be admitted or excluded, but whether experimental demonstration should be excluded, for teaching purposes, altogether.

When I was teaching systematic physiology, as I did teach it in a medical school or many years, I abstained for a long period from the direct experimental method; it was not expected of me, and it was not done. I found no difficulty, and my classes, I may say, worked satisfactorily. The students had the credit of becoming good physiologists, and I am sure there was nothing shirked. In the later part of my time I followed, occasionally, the plan of making a few experiments in the way of demonstration ; and, although these were rendered painless, I could not fail to observe that the innovation was not the success that was expected. In the first place, it led to much loss of time, so that other subjects than those under experimental survey were omitted. In the next place, owing to different ways in which pupils watched the processes of nature that were laid before them, there was often a difference of opinion in respect to what appeared to me matter of fact. If, for example, several students were at one and the same time looking at the motions of the auricles and ventricles of the heart of a narcotised animal, two of them rarely saw the same order of the phenomena, and at first most of them, like Harvey himself, were perplexed. One saw what the other did not see ; and, although they might all defer to my explanation, they were not all satisfied. I remember a very clever student saying to me on this very point, " It is not so clear, sir, as you made the fœtal circulation," which meant that a much more obscure subject, one that could not possibly have been demonstrated by experiment, had been made, to his mind, more distinctive. Thirdly, I found that the experimental introduction disturbed the order of the course; it led many students to concentrate their attention on particular subjects, as if they ought to be individual investigators of such subjects, instead of being students of the whole field; and I am sure it incited some ambitious students, who had no special genius or talent for original inquiry, to give up really useful learning which they could follow, for a childish ambition to become great men in the new line that had been opened to them.

Intellectually, I do not think my classes were assisted, in the main, by the experimental demonstration. I am sure it limited my sphere of usefulness, by leading me, in the limited space of time at my command, to omit some parts of physiology of a simpler, less controversial, and more useful kind. I am bound to say, too, that, morally,

I do not recall the effect as producing all that
could be wished. It is fair to say that I never
knew a student begin to study experimentally
on living animals on his own account, and I feel
certain that outside statements about students'
experimentalisms are, if not absolutely untrue,
rank exaggerations; but what I found was that
natures were so different no two approached
experiment with the same mind. Some were
so sensitive they could not look on with any
benefit of instruction; they stood behind, seeming
to see, in order not to be peculiar, but in truth
not understanding anything properly, and count-
ing the time when all would be over; others,
on the contrary, were enthusiastic in their
manner, were anxious to assist, were ready with
questions foolish or wise, and sometimes saw on
their own account what really was not shown,
the imagination playing too keen a part; a few
looked on impartially, and gained something;
but even amongst them were divisions into the
satisfied, the doubtful, and the dissatisfied. This
was the kind of chaos which I witnessed in
demonstration of experiments to students. Of
the hardening process of which I have heard

I saw nothing. I now and then saw nervous
students become indifferent, not actually callous,
as to the effects observed ; but this was the
worst, and I do not accept that natures are
hardened by experiment. The sensitive remain
sensitive still, although for appearance' sake they
may assume some bravado ; and the insensitive
remain as they were, cool, critical, and inquisitive.
They who are affected unfavourably are the
weakly ambitious, who think they are born to
discover, and, destitute of insight, raise themselves upon their feeble toes, and try to look
over the heads of their fellows towards positions
they are never fitted to occupy, even if by some
fluke they win them.

To be plain, I soon gave up experiments in
my classes, not from any sentiment, but because
I got on better without them. The omission
enabled me so to teach as to carry the student
in two six-month courses through the larger
domain of physiology that was useful to him.
I did not omit the facts derived from experiment,
I did not omit the report of my own experimental endeavours ; but I omitted repeating, for
the mere sake of demonstrating, what seemed

to have been proved, and I avoided testing
the unproved, on the ground that the audience
was composed of students, not of critical
compeers.

In like manner, in public courses on physiology, I have omitted painful experiments on
living animals with the good results of avoiding
all the controversy incident to experiment, of
saving time, and of the freedom from everything that created a disturbance in the minds
of the learners.

I am inclined, for the reasons assigned above,
to object to the demonstration of experiments
on living animals before students of physiology,
and especially to the performance of painful
experiments.    Were I a student, I should,
with my present knowledge, much prefer for a
teacher a physiologist who was well versed in the
wide field of physiology, and who would carefully
lead me over the course of his knowledge, than one
who had experimentally studied a few particular
subjects, and was enthusiastically anxious to illus-
trate his researches by experimented methods he
himself had acquired.    The temptation no doubt
is great to teach what is thought to be original

observation, especially if the results have been from experiment, but it is a dangerous practice to follow, or to make it the chief object of professional labour in the medical or general class-room.

The case is somewhat different when the representatives of a class are themselves duly qualified men or women, who are anxious to profit from the evidence of new demonstrations. But the effect is not invariably satisfactory even then. For several years I delivered courses of this kind, which were well attended, but I was not satisfied with the final results. There are always some, often many, who express distaste ; there are some who question results as evidences bearing on, for, or against, particular hypotheses or theories ; while the few who express conviction are not necessarily the strongest or most desirable of disciples. Were I again to deliver a course of physiological lectures to qualified hearers, I should make the experimental demonstrations on living animals as few and far between as was compatible with duty. They would be exceptional of exceptional, and painless from beginning to end.

# CHAPTER VIII.

## EXPERIMENT UNDER LEGAL ENACTMENT.

QUESTION 8.—" *Considering the educational influence of Law upon any community, how far should Law intervene in relation to experimental research ?* "

I HAVE ventured to differ *in toto* from the first, with those enthusiasts who have forced the present system of legislation in England for regulating experiments on lower animals and for the licensing of experimentalists. The Act is nugatory, for of what use can a prohibitory Act be which a physiologist can laugh at by the simple process of crossing to Boulogne to pursue his investigations ? If the prevention of cruelty be the object, the act is also worthless ; because it only increases the supposed cruelty in countries where no such Act is in the way, and that by experimenters who are certainly not less curious, not less ambitious, and not more common-

sense than men of the English school. The
Act again is mischievous, in that it prevents men
of really original mind from working out valuable
original inquiries. Men like William Harvey,
Thomas Willis, John Hunter, or Wilson Philip,
could never have worked under it. Original
work is the product of original thought ; and
men of original thought and action, the exceptional men who alone are worth consideration
on great subjects, are naturally wise men who
have hearts retentive of their own. Men of
that stamp have flashes of thought, or long
labours of thought, peculiar to themselves ; and
for such men, to pray a political personage, who
may know nothing whatever on the question in
hand, for permission to hold a licence in the
pursuit of an inquiry, the nature of which
is original, and which must be carried out
with other men acting as a kind of scientific
police, is to stultify everything, and make it
appear that a free mind can labour in bonds.
New and great results under such a method are
impossible, except by some accident of observation, which the staunchest advocate of the
wretched Act would not bargain for, or admit as a

reason for legislative restriction. There are valid objections against the present licensing law; but they are minor when compared with the demoralising and degrading action of the law upon the noble profession of medicine. This law places the professors of medicine in the same position as the licensed publican, and for the same reason. The licensed publican is the vendor and dispenser of what is considered by the lawmaker as a necessary evil. It is necessary for the people to be able to buy for drink intoxicating, that is, literally expressed, toxic beverages; such is the argument. The buyers, it is assumed, were there no restriction of sale, would destroy themselves in great numbers; and the vendors, were there no restriction, would sell to that effect. Necessity for restriction is thought, therefore, to exist, and what is called necessity is made subservient to law. The law is that the vendor shall be held in check by a licence. The vendor may meet the necessity half, or some part of the, way to ruin, but he must be known, and be under observation. By the same argument experimental physiology is considered a necessity; but experimental physiology may go too far, and may shock the moral sense of the people ; therefore the experimentalist must be checked.   He must be made to meet necessity halfway, or some part of the way ; he must be known, watched, and licensed.   He is an educated man ; he is a gentleman ;  the lives of his fellow-men, in sickness, may be entrusted to him ; but in this matter of making experiments that might shock the moral sense of the fastidious he is not to be trusted ; he must be licensed.  His discoveries must be licensed ; his results bad as bad, or good as the best that a learned man can produce, must be licensed, placed under supervision, placed under a ban, held up to criticism, ridicule, disgust.

The insult to medicine is as unpardonable as the argument is illogical.  If the results to be obtained by experiment are even partly useful, they are widely useful.  If the effects of per-mitted experiments are a necessity, the effects of experiments not permitted may be the greater necessity, because they may be directed to the worst pains and evils to which mankind is sub-jected.  Moreover, every one who takes out a licence exposes himself to the suspicion that he

may be doing something that ought not to be
permitted, and that requires to be supervised by
persons who merely hold the law in their possession. In plain terms, the law is wrong. It is
wrong in those who enforce it ; it is wrong, very
wrong, in those who submit to take advantage
of it ; it tempts weak men to weak practices ;
increases the number of experimentalists ; makes
experiment all but useless, and does not limit
cruelty.

A true and earnest physiologist requires no
hand of the law to be held over him. If, in great
necessity, he feels conscientiously that he must
perform some particular experiment on a living
animal; if the *crux crucis* be before him, he
should perform the experiment without any
impediment, and let the result be his justifica-
tion. If he cannot stand firmly by his own act,
if he cannot reasonably justify it, then only
should the law intervene.

It is asked, "Considering the educational
influence of law, how far should law intervene
in relation to experimental research ?" The
answer is given above ; but I would add that
in the present system law has no educational

influence whatever. There are now in England more experimentalists and more experiments than there were before the Act was passed, with a smaller number of important results.

In brief, the Act has not prevented, but has legalised and extended experiment by enabling almost any one who will condescend to solicit the favour to become an experimentalist. In this manner it has made experimental work a kind of speciality or profession of itself, with the effect that investigations are not now instituted to solve, simply, important questions relating to successful medicine or surgery, but for the settlement of subjects in disputation, and of questions that are curious rather than necessary and applicable to practice. Whether there could be an educational law is an open question. I know that good physiology is much impeded and injured by the present law, and I know that the objects, for which that law was rashly enforced, with little knowledge and less wisdom, have not been obtained ; but if there must be law, let it have something in it that shall be sensible and preservative of rational freedom. The most sensitive of the public may be sure that no good can follow the act of degrading learned men and forcing them to accept what must be a continual worry and insult to every honourable and independent mind.

No one without prejudice, whether he or she be for or against experimentation, can wish, on reflection, to maintain the present law—a law false in principle all round. A physiologist, for some particular purpose, asks for a licence of a minister of the State. How is a states-man, whatever his good qualities be, able to test the reasonableness, the necessity, or the origin-ality, of a physiological research? He may refer the matter to other physiologists, but what sub-stantial improvement is there? The committee of one or more to whom the matter is referred may be an able committee; but how can one mind see efficiently into that of another, or seeing, be, as is absolutely required for the sake of honesty, free of all previous conjectures, feelings, and self-beliefs? In Harvey's day there probably was not a single professor who would not have ignored his theory on the mere paper proposition of it in advance, although many of his views had actually been foreshadowed; while he himself would have been under the apprehension that if he disclosed his project some unscrupulous and less competent man would get hold of it and forestall him in discovery. The method is a simple absurdity, and would not be entertained for a moment in any other branch of intellectual inquiry. Imagine a chemist, astronomer, engineer, driven to pursue original inquiry under the same conditions! All would alike fail. Better it were to repress experimental physiology altogether than to tie it to such weak and ignoble principles of action.

Without presuming, for a moment, to speak the thoughts of any other physiologist, the only law I could commend to consideration would be a new one altogether. I would not object to have it enacted that all vital experiment for teaching purposes should be prohibited; but I would leave properly qualified men free to pursue their investigations on the plan that in promoting every new discovery or theory founded on experiment, they should, if necessary, furnish the details by which they arrived at their conclusions, leaving it open for the law to decide, on the instance of competent authority, whether what was done was justified by the results. This might leave ground for public trial, and for the hearing of both sides, under which the public would very soon make an appreciative judgment that would be just to its own interests, to the progress of science, and to the inferior animal kingdom. To put the matter in a nutshell, I would place the responsibility of all experimentation not alone on the experimentalist, but equally, on that community for whose benefit his unthankful services are performed—services which, unless they are useful, directly or prospectively, to the community and to science, would be illegal altogether.

A new law, founded on the principle of dividing responsibility instead of restricting the freedom of a few particular men, might possibly lead at first to legal action. But this would be of short duration. The Judges would soon produce precedents which would have the force of law ; and after a few trials what could and what could not be done experimentally would be clearly defined. In the end, the conditions that would favour or prevent experimentation would be known, and things would find their proper level on public issues, which is the great desideratum. The public would quickly feel that the true and safe physiologist, although he might have to do what to himself was repellent, was working for the public benefit; and he would be saved the reproach, distrust, and dislike under which he now pursues his wearisome and despised career—a career which many, in the calm of private life, shrink from as decisively as its fiercest opponent.

Nothing in our age has produced so much rank insincerity as the physiological *crux crucis*. The violent, unmeasured, and open wrath of the most severe of the anti-vivisectors is often better than the concealed hatred of those who say nothing, or, under the idea of being one of the majority, say something exculpatory of experimentation in equivocal terms. Persons sometimes, under the impulse of fear, excuse or tolerate experiment in affliction, if it bear on a special case in which they are, for the moment, anxious. But, the anxiety over, they are no longer apologetical about it. If experimental practice be not ˙ satisfactory they are more incensed than ever; while, if it seem to succeed they forget the benefit, find reasons to doubt whether the practice, after all, had useful influence, and, if they hear of recoveries of a similar nature in which experiment has not been invoked, think they have been sold in distress, and cajoled in an hour when their faculties were dimmed by their affections ; a dismal conviction.

Perhaps some will conclude, from what has now been said, that special Acts of Parliament on experimental work are useless altogether, and that the common law is all-sufficient. That is an extreme view, and it certainly would not satisfy a large and active number of the educated community, who fear that to give absolute freedom to too warm an enthusiasm for experiment might lead to the perpetration of certain senseless and most painful continental exhibitions, which, as insanities of enthusiasm, every sober-minded man and woman condemns. If legal protection to animals is necessary, it should be a protection extending through all classes, not alone through the one class that is the mere instrument of the national will. In old Egypt it is said that the operator, who with a sharp

stone made the first incision into the dead body
that was about to be embalmed, had to "cut
and run," so fiercely was he assailed as he performed the first necessary part of the ceremony.
The modern experimentalist is much in the same
position—a position in every sense, and to every
one, unjust. If what he does is useful his act
ought to be by the approval of all; and in so far
as he is useful he ought to be held in honour; but
the responsibility of the act should be distributed.
In great trials where human life is at stake it is
so. The ruling powers instruct their experts to
test, on the lower animal, the effects of certain
poisons or suspected poisonous substances. The
experts make their researches, give evidence
upon them before the judge, the jury, and the
public, and no one feels that wrong has been done,
though many have no confidence in the verdict.
Even if it has been necessary to inflict pain by
the experimentation, no fault attaches to the
experimenter, because he is coming before the
public as a defender of its rights. He has performed a disagreeable duty for the public, and is
treated by it with proper consideration, against
which the criticisms of the bar itself, if the

research be honest and explanatory, are of no lasting character. On such occasions the experimentalist is in his true position, the position in which he would always stand, were the law I have ventured to recommend at once corrective, protective, and responsible.

## CONCLUSIONS.

To sum up: I would offer on the question put to me—Question 8—the following suggestions :—

1. Experiments on living animals should only be performed for a definite object, which every one can understand when the nature of it is faithfully and clearly explained.

2. Experiments on living animals should not be a matter of mere demonstration to students in the schools of physiology.

3. If demonstrations on living animals become necessary in order to settle for good some important vital subject, they should always be performed by duly qualified men and accepted by fully qualified judges. Otherwise they should not be considered as legal acts.

4. All experiments on living animals should

II

be conducted, as far as possible, painlessly ; subject to the legal responsibility before described.

5. All experiments performed for the public service, as in trials for injury or murder, should be conducted by a properly qualified commission, acting under the direct protection of the law, and conveying in full its results to the public in the usual form of report.

# CHAPTER IX.

## A SCHOOL OF PREVENTIVE MEDICINE.

QUESTION 9.—" *Can you lay down an outline of an Institute of Preventive Medicine, where the objects and methods of research shall be in accordance with the terms and the spirit of the Leigh-Browne Endowment ?*"

IF I understand it correctly, the great object of the Leigh-Browne endowment is to establish an institution for the study of biology, in which advanced views for the prevention of disease shall be taught, and in which experimentation and observation in all directions shall be considered as the bases of research, but from which painful experimentation shall be altogether excluded. If this be correct, there is no practical difficulty in the way, since nothing is required for such a school except a solid foundation, and sufficient means for the endowment of competent men as teachers or professors. In the organisation of the plan the question of position need give no trouble. For every reason London should be the centre—firstly, because it is really the vital centre, the heart of the living world at the present moment; secondly, because it possesses the greatest resources for learners; and lastly, because in its museums and gardens it holds the most extensive and choicest treasures of animal nature living or dead.

I am of opinion that the foundation should be a school or university, not an institute. If an institute were founded it would be a centre of discussion, as distinct from learning and teaching. That is not what is wanted at the present time. We have now in our possession the accumulated original learning of many centuries, and we want to have that systematically distributed to the young, so that they may advance upon it, and in no direction is this event more felt than in preventive medicine.

The sections or classes for the teaching of such learning should be :—

1. *Biology.* Including the natural history of plants, animals, and men.

2. *Physiology.* Including anatomy and physiology, human and comparative.

3. *Psychology.* Including the relationships of mental laws and developments to the physical.

4. *Construction.* Including sanitary building, engineering, and decoration.

5. *Chemistry.* Including organic and inorganic chemistry.

6. *Hygiene.* Including laws of health in plants, animals, and men, vital statistics, and anthropology.

That six professors could be found who would systematically conduct these sections or classes admits of no distrust or doubt ; and, with competent assistants for departments, most complete courses could be carried out in fixed periods or terms extending over the year according to the direction of a board of management, or committee of the Trust, which would from the first be established, and to which the professors, as a senate, could act in all matters of an educational character.

In suggesting this plan of a school or university of preventive medicine, I am thinking of a foundation ; but it would be unwise to consider it as a teaching establishment and no more. The professors in it should have every encouragement for original research ; each professor and teacher should consider it his duty to learn as well as to teach, and should be encouraged to become a master in his own department.

The school would not be, in any sense, a rival of any medical school, nor of any society in which the reading of papers or debates were conducted. A voluntary society might be attached to it, and the buildings of it might be the home of other societies ; but these, if permitted, would be independent bodies, having their own work and bearing their own responsibilities, pecuniary and intellectual.

The secondary object of the school would be to educate students of both sexes in the laws of life and the prevention of disease. It would proceed on the idea that the prevention of disease must depend on individual knowledge and practice. As a university or school of preventive medicine it should be open to all students ; as a university of health, in which infliction of pain would play no part for discovery, and in which all that was discovered and taught would be for the health of the mind as well as the body; its aim being the inculcation of physical science and physical well-being, adapted in an inexpensive form, to the intelligence of the people, so as to bring it within the reach of all who had been through a Board School.

At first the university would have to stand supported by the Trust, by other aids that might be left to it, and by very moderate fees derived from students who were able and willing to pay for the instruction afforded. In time it might, and I have no doubt would, have State recognition for its chairs, its libraries, its museums, and, ultimately, its degrees.

In the outline of subjects, as divided into classes or sections, I have referred to the heads only of the various classes. It will be seen how rich a field for the observation of nature, how wide a scope for both teacher and student, would be opened under each section. I suspect that many will say that each department is too broad, and that in this day of "divide, divide, divide," there should be three times the number of departments and professors. I venture to think differently. In my view specialism has

been carried into imbecility. Everybody now
who is incapable of grasping a great department
of science thinks it is his business or mission to
take up a speciality, and by the minute character
of his own littleness to construct a new science
of the puny order. All this requires to be
reformed throughout the sciences, but in none
so much as in preventive medicine, where, as
at the present moment, particular studies, such
as bacteriology, absorb all the rest. In a
proper school no branch of science, however
small, should be neglected : that is quite true,
for "there are mites in science as well as in
charity"; but minor branches of great subjects
are best taught when the mind of the teacher
is broad enough to place the minor in its true
position while still expounding it according to
its worth. In the new university or school it
would be the duty of the professors to give an
extended education of health science altogether ;
not to waste time in a branch of science which
cannot be taught soundly until it is accorded
its right position as a part, and no more, of the
broader science from which it springs.

From a long experience as a teacher of

physiology and of public health, I am convinced that a school or university of preventive medicine on the lines laid down would fill an 'important want, and become exceedingly popular. It would tend to make every man and woman a sanitarian, and would help to bring the principles of health into every home. That would be its direct and practical utility, but it would do more ; it would instil an exalted comprehension of the designs of nature, of natural laws, of the advantages of following those laws, and of the danger and 'folly of setting them at ignorant defiance. It would thus sustain a natural religion with science as the groundwork of fact, belief resting on positive knowledge. ·

Some will perhaps think that the scheme proposed is too comprehensive. I reply that comprehensiveness is the necessity most urgently demanded at this crisis when science is without philosophy. We are growing like Babel. One man does not understand any language, any science, any art, except his own. Men make a boast of their own narrowness, and raise the *ne ultra crepidam* motto into blustering absurdity. We must learn and teach more, bring our followers into closer consonance with nature, in which task they would find no more difficulty than in the minute study of minor details; while their own minds expanding under the instruction, would discover the unity of design in nature and how to apply wisdom to knowledge.

The end would be, after all, the accomplishment of the great aim ; the development of the health of the people ; the art of preventive medicine without inflicting pain on any living thing, and without the effect of injuring physiology ; but, with the effect of hastening a time when the humanest treatment of the inferior animal world would be one of the brightest features of humanity, with health as the standard.

Printed by Hazell, Watson, & Viney, Ld., London and Aylesbury.

www.ingramcontent.com/pod-product-compliance
Lightning Source LLC
Chambersburg PA
CBHW021805190326
41518CB00007B/461